AF352986

Case Studies
3

Investigation of the Chirajara Bridge Collapse

authored by
Prof. Christos T. Georgakis
Prof. Yozo Fujino
Siegfried Hopf
Klaus H. Ostenfeld
S. Eilif Svensson

with the assistance of
Lasse Hindhede, Martin Juul Krogstrup,
Michael Müller, Francisco Subirá,
Gregor Fischer and Thomas Rumpelt

International Association for Bridge and Structural Engineering (IABSE)

Copyright © 2022 by

International Association for Bridge and Structural Engineering

All rights reserved. No part of this book may be reproduced in any form or by any means, electronic or mechanical, including photocopying, recording, or by any information storage and retrieval system, without permission in writing from the publisher.

ISBN: 978-3-85748-185-7 (print)
eISBN: 978-3-85748-186-4 (PDF), 978-3-85748-187-1 (ePUB)
DOI: https://doi.org/10.2749/cs003

Publisher

IABSE
Jungholzstrasse 28
8050 Zürich
Switzerland

Phone: Int. +41-43-443 9765
E-mail: secretariat@iabse.org
Web: www.iabse.org

This book was produced in cooperation with Structurae, Dresdener Str. 110, Berlin, Germany (https://structurae.net).

Copyediting: Jens Völker, Hilka Rogers, Nicolas Janberg
Layout & typesetting: Florian Hawemann

Preface

On January 15, 2018, at 11:49, the west pylon (B) of the cable-stayed Chirajara Bridge collapsed during construction of the bridge girder. The crossing is located approximately 20 km NW of Villavicencio, Colombia. The collapse led to the complete destruction of Pylon B, together with the erected span of the bridge girder. Authorities reported nine fatalities resulting from the collapse.

Shortly thereafter, the project insurer QBE Segures (Colombia) commissioned an independent investigation into the collapse of the bridge, through loss adjusters ONC Adjusters (Bogotá, Colombia). An international team of bridge engineering experts was assembled to undertake the investigation, with Professor Christos T. Georgakis of Brincker & Georgakis, Denmark as chair of the team. The other members of the team are Sven Eilif Svensson of ES-Consult and Klaus H. Ostenfeld of KHO-Consult in Denmark, Siegfried Hopf of Leonhardt, Andrä und Partner in Germany and Professor Yozo Fujino of Yokohama National University in Japan.

On May 30, 2018, the team issued a brief interim report on the causes of the collapse of Pylon B. Demolition of the standing east pylon (C) was recommended in the brief interim report, as it was nearly identical to Pylon B and subject to the same deficiencies as identified for Pylon B in this report. The expert team was informed that the demolition of Pylon C and the remaining parts of the superstructure was carried out on July 11, 2018.

On August 4, 2018, an extended interim report was issued, elaborating the findings presented in the brief interim report concerning the detailed failure mechanism of Pylon B. In addition, the extended interim report presented a general assessment of the overall bridge design and the corresponding design flaws throughout the bridge, as well as a brief geotechnical assessment.

All findings of the Brief Interim Report of May 30, 2018, and the Extended Interim Report of August 4, 2018, are presented and expanded upon in this book. The final findings of the team on the detailed investigation into the failure mechanism of Pylon B are reported in addition to a general assessment of the bridge design, the materials used for construction, and geotechnical aspects, whilst also presenting the observations made during site visits and interviews with all relevant parties.

The investigation has led to the following main conclusions:

- A detailed design check of the bridge, as defined in the engineering drawings and reports provided to the team, revealed several important design flaws.

Design flaws include, but are not limited to, the design of the stay-force transfer mechanism in the pylon head, the bearing layout and abutment design, the aerodynamics of the girder, and the lack of appropriate levels of ductility and lateral restraint at the pylon knees.

- After a detailed investigation of the possible causes of collapse, it was concluded that Pylon B would fail during construction at loads well below the design loads.

  The primary deficiency was the lack of lateral tie capacity between the pylon knees. Thus, it is concluded that the design is faulty. The faulty design alone would have led to the collapse of the pylon, irrespective of any other potential contributory factors. Furthermore, the sudden nature of the collapse, without warning, can be attributed to the non-ductile (brittle) design of the diaphragm between the lower legs.

- Deficiencies in the caisson construction may have led to the need for future retrofit or strengthening but did not ultimately contribute to the collapse of the bridge.

- It is recommended for projects of this scale and importance that an independent Category III design check be undertaken. Furthermore, internationally accepted best practices for the procurement of projects should be adopted, an ISO 9000 or similar system should be implemented and followed, realistic time schedules should be pursued, and robust and appropriate communication between all parties should be established.

  Appropriate degrees of structural robustness, redundancy, and ductility should always be ensured.

The Authors would like to thank IABSE Bulletin Board, and especially the Vice Chair and Chief Reviewer for this Case Study, Dr. Fabrizio Palimisano (Italy), and the Reviewers from IABSE Task Group 5.1 Forensic Engineering, John Duntemann (US) and Laurent Rus (Spain), for their valuable inputs to prepare this document for publication in a Case Study format for IABSE.

Prof. Dr. Christos T. Georgakis
On behalf of all authors

Table of Contents

Chapter 3
Status of Construction and Loads Immediately Prior to Collapse 33

Chapter 4
Description of the Collapse 41

Chapter 5
Site Visits 49

Chapter 6
Interviews 77

An Overview of the Forensic Investigation Process

John F. Duntemann, Senior Principal, Wiss, Janney, Elstner Associates, Chair, IABSE Task Group 5.1. Forensic Structural Engineering

Introduction

Engineering investigation of the causes of structural failures of buildings, bridges, and other constructed facilities, as well as rendering opinions as to the cause(s) of the failures, is a field of practice often referred to as *forensic structural engineering.*

The process of forensic structural engineering investigation is different from conventional structural engineering design. An engineer performing a forensic structural engineering investigation generally has the benefit of the evidence, which, if well-documented and correctly analyzed, can explain how and why the failure or collapse occurred.

The investigation of structural failures generally consists of the following tasks:

- First response and preliminary assessment
- Development of investigation plans and protocols
- Fact gathering and document review
- Engineering analyses to determine the cause(s) and responsibilities
- Reporting on the findings of the investigation
- Recommendations on how to avoid repeating the same mistakes

The investigation of the Chirajara Bridge Collapse as reported in this book includes many of the elements described above.

Fact Gathering

The collapse of the cable-stayed Chirajara Bridge near Villavicencio, Columbia occurred during construction on January 15, 2018. There were reportedly nine fatalities due to the collapse.

The experts in charge of the independent investigation were retained by the project insurer to determine the cause of the collapse. They were allowed access to the site on two separate occasions, March 20, 2018, and March 22, 2018. The first site visit included access to the collapsed Pylon B and Abutments A, and the remaining (uncollapsed) Pylon C and Abutment D.

Before the Collapse

The status of construction immediately prior to the collapse was determined from various videos of the bridge that were recorded on the day of the collapse. No unusual construction work was being performed at the time of the collapse. A review of the available wind and temperature data prior to collapse also did not indicate any unusual conditions or extreme changes. While three minor seismic events were recorded on the day of the collapse within a radius of 200 km from the bridge site, it was concluded that these seismic events did not contribute to the collapse.

During and After the Collapse

Video footage of the collapse and 55 minutes prior to the collapse was available for review by the investigators. The video footage indicated no significant wind conditions or ground movements at the time of the collapse. The duration of the collapse was about 7 seconds which started with slacking of the shortest main span cables and ultimately the collapse of Pylon B.

Subsequent examination of the debris pile indicated that the structural components of the pylon ended up almost directly below their original (pre-collapse) position on the longitudinal axis of the bridge. The investigators observed that both the southern and northern lower pylon legs had separated from the diaphragm along the entire length of the legs by rupturing all the horizontal reinforcement at the inner face of the lower pylon leg. The pylon head was observed to be largely intact, except for damage by concrete crushing and ruptured reinforcement where the head was connected to the upper pylon legs.

Interviews

The general contractor, the design engineer, the site engineer, and the cable supplier were all interviewed as part of this investigation. These interviews provided important information and clarifications regarding the design of the bridge and the construction sequence and methods.

Engineering Analyses

Design Review

A review of the bridge design indicated several significant design deficiencies including a severe lack of bracing tie capacity of the link slab and diaphragm for design loads, service loads, and during erection; the bearings did not have sufficient capacity to withstand the calculated shear forces and deformations during construction and for the service stage design loads; the anchor beam assembly was inadequately designed for service loads; the pylon head was inadequately designed for the calculated longitudinal forces; the transition beam and corresponding connections were not designed with sufficient tension capacity; the longitudinal edge beams of the girder were inadequately designed; and the girder concrete slab was not sufficiently reinforced in large negative moment areas.

Geotechnical Review

A review of the original geotechnical investigation and the foundation design was also performed as part of the forensic investigation. The investigators concluded that the failure of the pylon was not caused by unforeseen settlements, lateral movement, or induced loads through failure of the tieback anchors, earthquakes, or other geotechnical factors.

Material Sampling and Tests

Material samples were taken from the collapsed bridge to determine their condition and material properties. These samples included concrete cores, steel reinforcement, steel tendons, and steel plates.

Finite Element Analysis

A detailed material and geometrical non-linear finite element analysis (FEA) of the collapsed bridge was performed to determine the collapse mechanism and the cause of the collapse. The purpose of the FEA was to analyze the actual structure and the actual loading at the time of the collapse. Thus, no load factors or capacity reduction factors were applied to the loads or the materials. The material properties were based upon the material testing mentioned above.

The FEA determined that the lateral connecting reinforcement between the pylon and diaphragm ruptured in the upper corner of the diaphragm after the external load caused deformations that exceeded the local ultimate deformation capacity. It was determined that the external load reached the level necessary for collapse three days after the last major construction activity, i.e., pouring of the concrete deck, and was attributed to several normally inconsequential factors or their combination. After the rupture of the upper lateral reinforcement, the connection between the diaphragm and the lower pylon legs failed in a non-ductile asymmetric manner, so that the south pylon leg detached from the diaphragm by an "unzipping" of the lateral reinforcement through the sequential rupturing of the whole diaphragm.

Reporting and Recommendations

A detailed review of the bridge design, as defined in the engineering drawings and reports provided to the investigators, revealed several important design flaws. The primary design deficiency was determined to be the inadequate lateral tie capacity between the knees of one of the two pylons. It was further determined that the sudden nature of the collapse was due to the non-ductile design of the diaphragm between the lower legs of the pylon. The investigators recommend that projects of this scale and importance are properly peer-reviewed and the procurement methods are improved.

Summary

The collapse investigation described in this book is a good example of a proper forensic engineering study. The information before, during, and after the collapse of the Chirajara Bridge is well documented and provides the basis for an accurate engineering analysis of the cause of the collapse. The subsequent bridge design review, the non-linear finite element modeling, and material sampling and testing served to correlate the information, or evidence, with the collapse mechanism and cause of the failure. The illustrations in Chapter 9 of the book are particularly illustrative when compared with the condition of the remaining east pylon that did not collapse. Finally, the recommendations, or lessons learned, in the book provide useful insights on how this collapse could have been avoided and will hopefully be referenced when constructing similar structures in the future.

Chapter 1

Description of the Structure

The Chirajara Bridge was situated in the Guayabetal municipality of Cundinamarca Department, nearly 20 km WNW of Villavicencio, capital of Meta Department, and approximately 60 km SE of Bogotá. Located at kilometer mark 64 of the National Route 40, the Chirajara Bridge formed part of a large infrastructure project of Agencia Nacional de la Infraestructura (Public Authority of the Colombian Government). The plan included a dual carriageway between Bogotá and the Eastern Plains of Colombia.

Excavation for the foundations of the bridge was initiated by the former constructor TRADECO in 2014. Later, in 2016, TRADECO's contract was terminated, whereupon the constructor, GISAICO, was appointed by the main contractor, CONINVIAL, to finalize

Fig. 1.1 Pylon C, identical to Pylon B, of the cable-stayed Chirajara Bridge

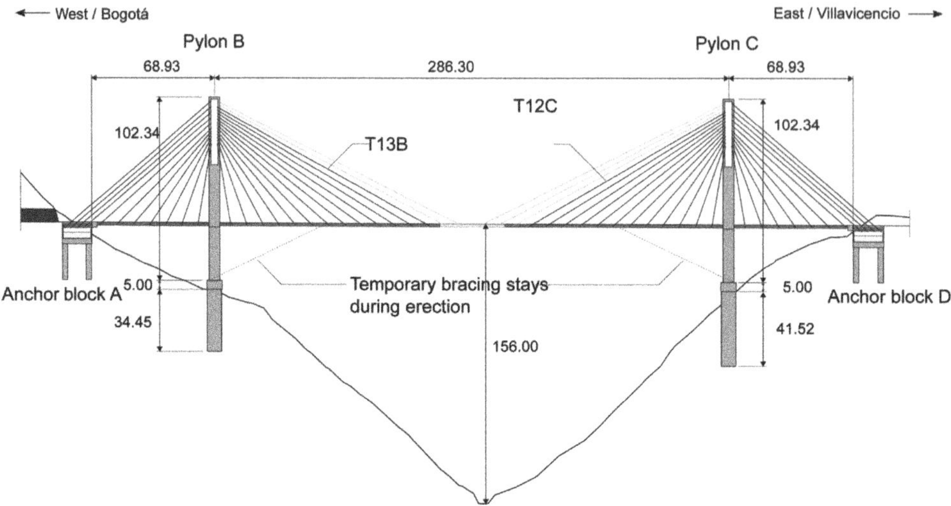

Fig. 1.2   General dimensions of the Chirajara Bridge. Members of the main span (cable + girder) shown in light grey shading had not yet been erected at the time of the collapse of Pylon B

the detailed design and continue the construction work. Scheduled to be completed in the first half of 2018, the Chirajara Bridge was designed to support two vehicular traffic lanes and two pedestrian walkways.

1.1 Bridge Geometry

The main dimensions of the cable-stayed bridge are presented in *Fig 1.2*. The stay cable system can be characterized as a double multi-cable semi-fan system with four parallel sets of anchor cables, which are the last four cables of the back-span. Apart from these anchor cables, each of the side spans comprised 7 sets of stay cables. Both main-span fans were to consist of 15 sets of cables each. At the time of the Pylon B collapse, 13 sets of cables were installed on the west cantilever for the main span, while only 12 sets of stay cables were installed on the east cantilever for the main span.

The main span of ~286 m and two side spans each of ~69 m add up to the intended total span of 424 m, spanning above the 156 m deep valley. The bridge girder was supported by two main pylons, Pylon B (West / Bogotá side) and Pylon C (East / Villavicencio side). Apart from the lengths of the mono-caissons Pylons B and C were identical, with a height of 102.34 m. The longitudinal profile of the bridge was rectilinear with a descending slope of 0.32 % towards the east. Thus, the level of Pylon C was 0.91 m below that of Pylon B.

The mono-caisson of Pylon B was ~34.5 m long and 8 m in diameter, with a 5 m deep solid concrete caisson cap on its top. The mono-caisson of Pylon C had a length of ~41.5 m.

Anchor blocks were located at each abutment, both being supported by four 20 m long caissons with a diameter of 3 m.

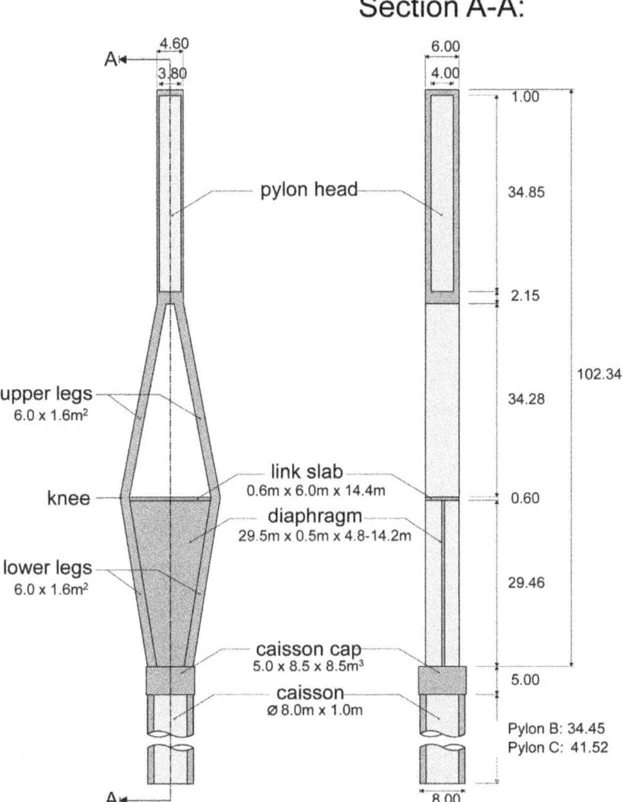

Fig. 1.3 Main dimensions of pylons

1.2 Structural Elements

1.2.1 Pylons

Both pylons were constructed of reinforced concrete, extending to a height of 102.34 m, measured from the top of the caisson cap. The overall widths of the pylons were 8.00 m, 17.62 m, and 4.60 m respectively at the bottom, knee, and top level of the pylon diamond. In addition to the foundation, the pylons consisted of five distinct structural components (see *Fig 1.3*):

- Lower legs
- Upper legs
- Diaphragm
- Link slab
- Pylon head

Together, the inclination of the upper and lower legs made up the diamond shape of the pylon. The width (longitudinally along bridge axis) and thickness of the legs were 6.0 m × 1.6 m. Longitudinal reinforcement in the lower and upper legs was provided by

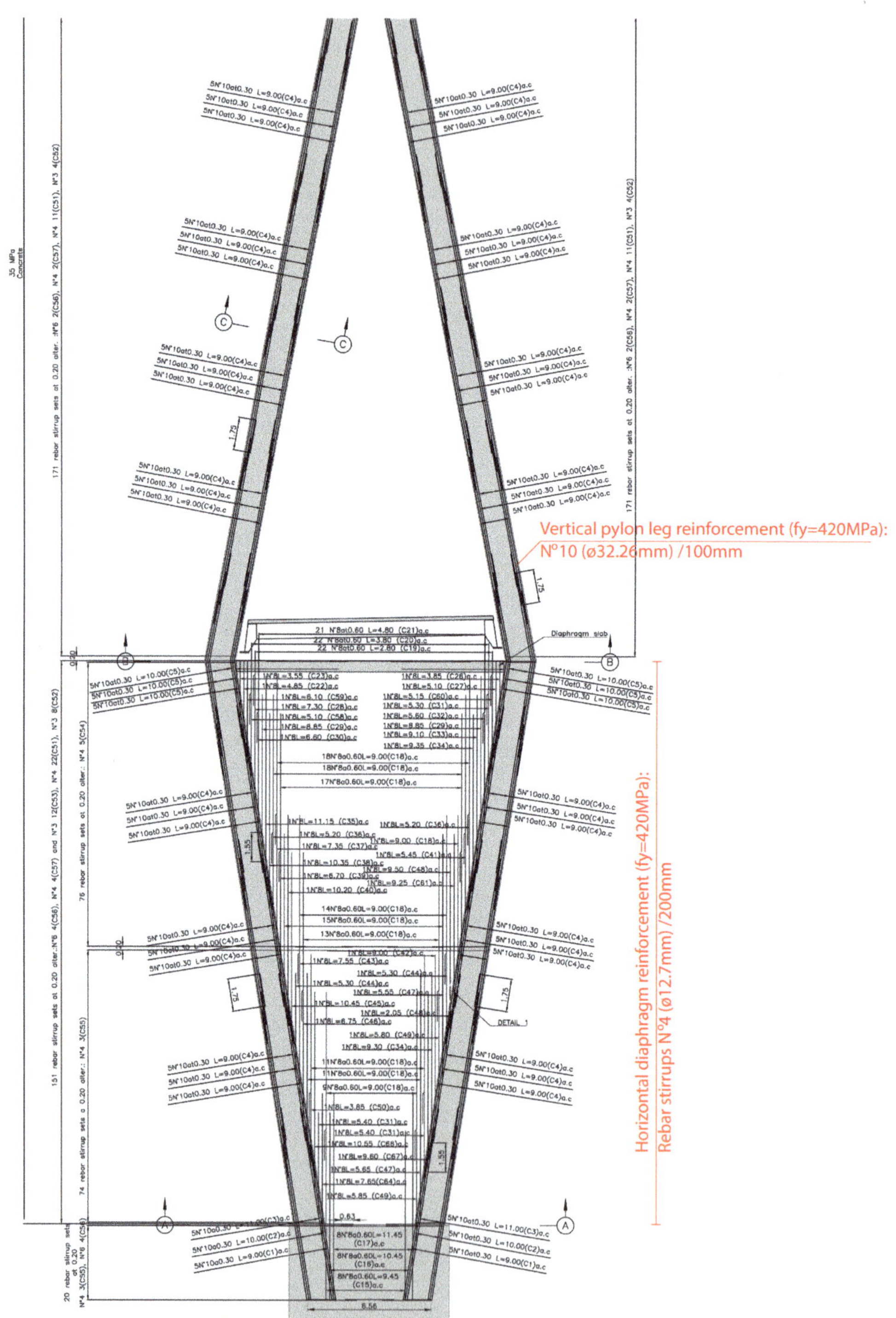

Fig. 1.4   Reinforcement in pylon legs and diaphragm.

N°10 bars (∅ 32.26 mm) distributed per 100 mm along all four outer faces of each leg (see *Fig 1.4*). The longitudinal reinforcement ratio was $\rho_{s,l}$ = 1.24 %. The inclination of the bottom legs with respect to vertical was 9.1°, whereas the top legs were inclined at 10.8° in the opposite direction.

A diaphragm with a thickness of 0.5 m was connecting the lower legs of each pylon. Vertical reinforcement was provided in the diaphragm as N°8 bars (∅ 25.4 mm) per 200 mm along both faces (see *Fig 1.5d*). Horizontal reinforcement was provided as N°4 (∅ 12.7 mm) overlapping stirrups of alternating length, spaced at 200 mm in the vertical direction (see *Fig 1.4*). These stirrups extended from the diaphragm and into the lower pylon legs. The vertical and horizontal reinforcement ratios were $\rho_{s,V}$ = 1.01 % and $\rho_{s,H}$ = 0.25 % respectively. The diaphragm was cast monolithically with the lower legs, thus constituting a vertical I-beam together with the lower legs, forming a vertical cantilever rising from the foundation caisson cap.

Situated on top of the diaphragm, between the pylon knees, was a 0.6 m thick link slab, extending the entire 6 m width of the pylon legs along the bridge axis. The link slab was post-tensioned in both longitudinal (along bridge axis) and transverse (knee to knee) directions. 14 tendons each consisting of 8 strands, 0.6" in diameter (nom. A_p = 140 mm²), were provided in the longitudinal direction of the bridge. 12 tendons each consisting of a single 0.6" strand were provided as ties between the knees of the pylon in transverse direction and anchored within the pylon legs. Horizontal reinforcement in the link slab was provided at the top and bottom faces by N°4 (∅ 12.7 mm) bars per 250 mm in both longitudinal and transverse directions. The reinforcement ended at the edge of the link slab. No reinforcement, other than the 12 single strands, was provided between the link slab and the pylon legs. The vertical reinforcement from the top part of the diaphragm extended into the link slab, bent as L-shaped hooks. The link slab was cast independently from either the diaphragm or the pylon legs, with a cold construction joint at the top of the diaphragm and against the pylon legs (see *Fig 1.5*).

The pylon head can be subdivided into three vertical segments, all with outer dimensions of 4.60 m × 6.00 m (W × L). The first segment, immediately above the upper legs and diamond, was a near solid rectangular section of dimensions 2.15 m × 4.60 m × 6.00 m (H × W × L) with a central access hole of 1.4 m × 4.0 m. The second and main pylon head segment was hollow, extending 34.85 m in height, with 1.0 m thick walls at the W/E faces and 0.4 m thick walls at the N/S faces. The third and top segment was another solid section of 1.0 m in height, with a 0.8 m × 0.8 m access hole. The inside of the main hollow segment contained a steel box with shear studs, which connected the steel box to the concrete through composite action. The box was made from individual rectangular plates of various dimensions, which were then welded together on site. Plates for the W/E faces had a thickness of 25 mm whereas the plates for N/S faces were 19 mm thick. Ducts in the concrete and pre-cut holes in the W/E face steel plates allowed for threading of the stay strands. The steel corbels for support of the stay anchors were welded to the steel plate after construction of the pylon head, just before the stay cables were installed. The stay cables were anchored and stressed from the inside of the steel box.

The foundation for each of the pylons consisted of a circular reinforced concrete mono-caisson, 8.0 m in diameter, with a wall thickness of 1.0 m. The caissons were con-

(a)

(b)

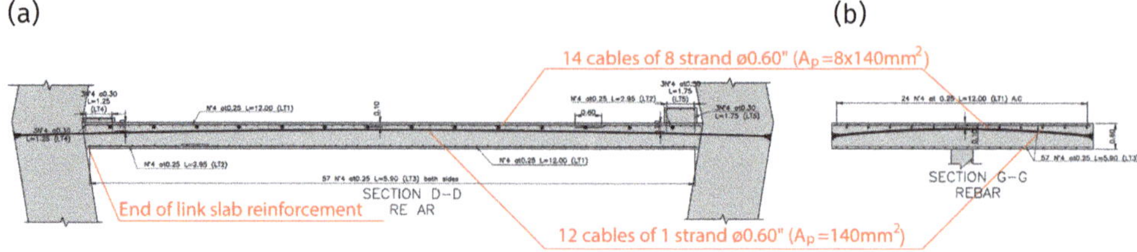

(c)

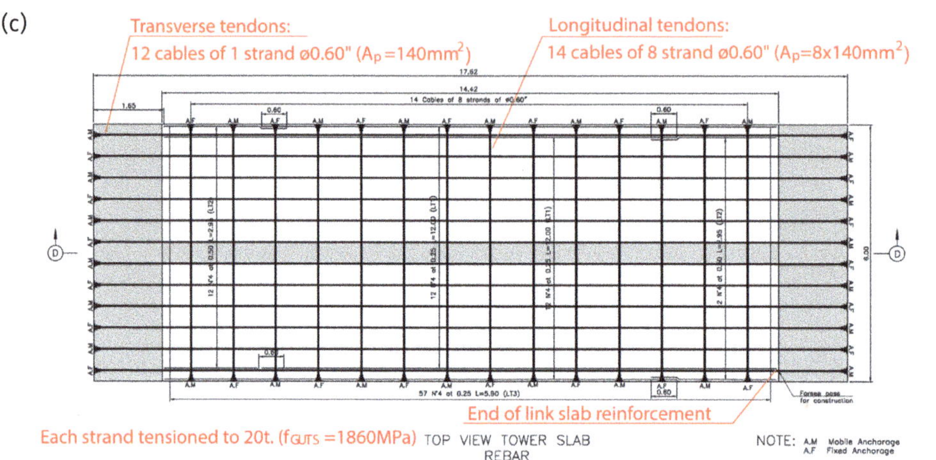

(d)

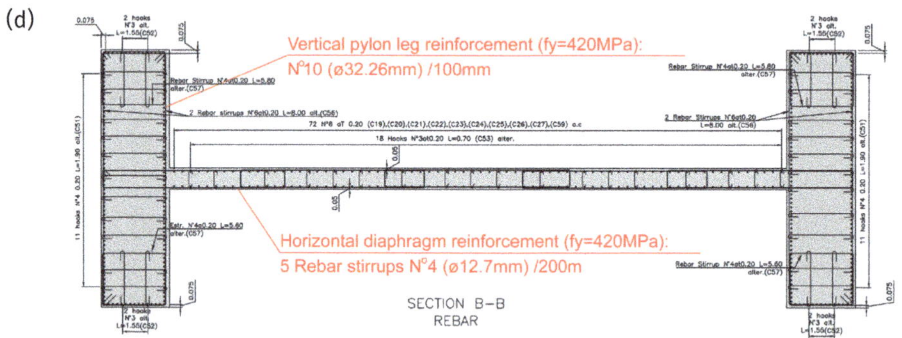

Fig. 1.5   Sections of pylon. a) Link slab, transverse section, b) Link slab, longitudinal section, c) Link slab, top view, d) Pylon leg and diaphragm, horizontal section.

structed step by step with segmented excavation followed by casting of 0.3 m × 0.6 m (W × H) concrete ring beams and the subsequent installation of post-tensioned tieback anchors, i.e., earth anchors. In between the concrete ring beams, a 0.2 m thick concrete outer wall was cast directly against the soil onto the ground. As this construction process reached the specified depth, the main caisson, a 0.8 m thick reinforced concrete wall, was cast in sections from the bottom and up, with the 0.2 m thick outer wall acting as formwork. The tieback anchors installed onto the ring beams were continued in ducts through the main caisson wall. On top of the caisson was a solid rectangular caisson cap with dimensions 5.0 m × 8.5 m × 8.5 m (H × W × L).

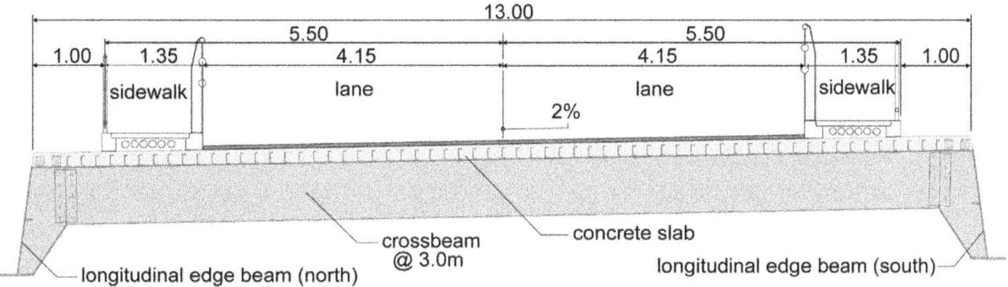

Fig. 1.6 Cross-sectional view of the composite bridge girder

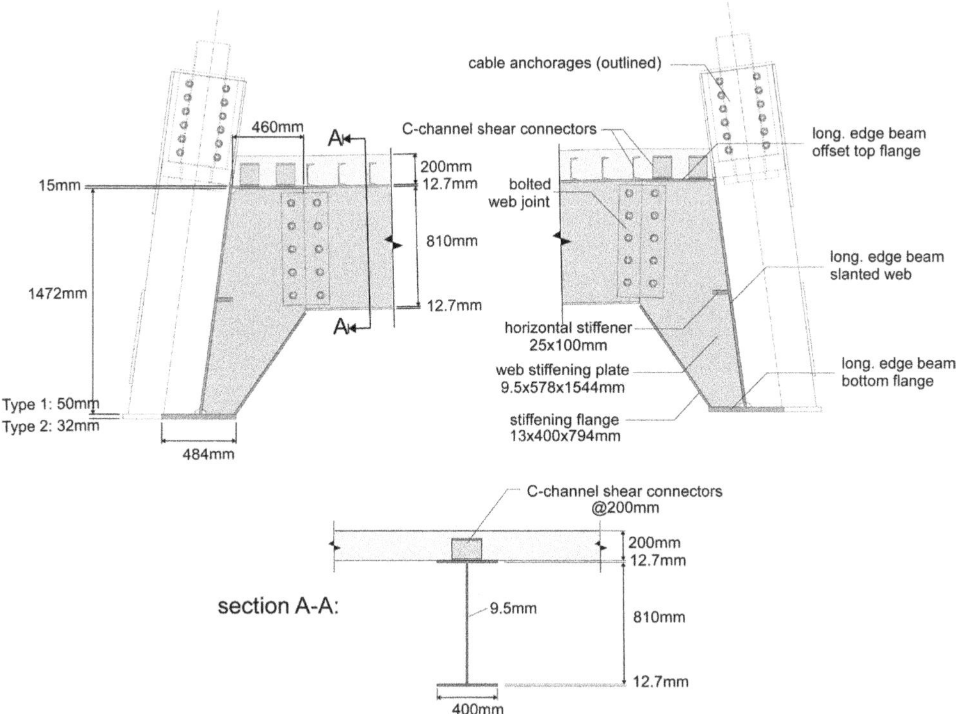

Fig. 1.7 Detailed view of longitudinal edge beams and connection to cross beams. (Cable anchorages outlined)

1.2.2 Girder

The bridge girder was a composite girder with longitudinal edge beams made of structural steel and a 200 mm concrete deck slab cast on top. The longitudinal edge beams were welded off site in a slanted I-shape with an offset top flange. Web stiffening plates were welded to the inside of the edge beams. Bolted to these stiffening plates were I-shaped steel cross beams, placed every 3 m, to support the concrete slab. The slab had a one-sided uniform 2 % slope in transverse direction, descending from the south towards the north (see *Fig 1.6* and *1.7*).

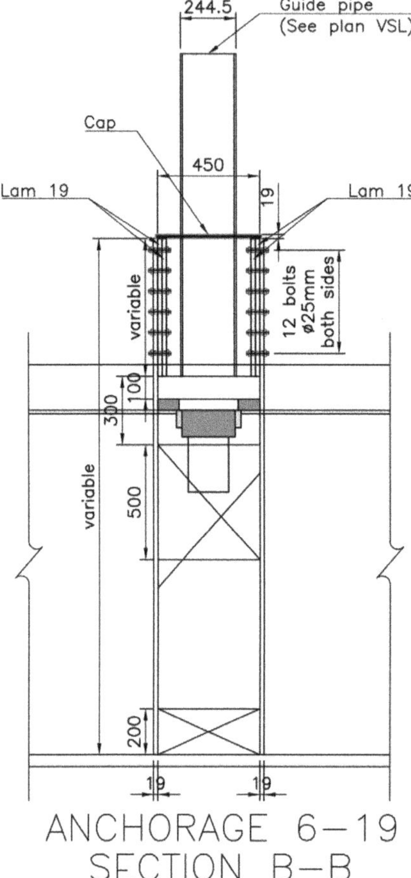

Fig. 1.8 Excerpt from design drawings showing the specification of the stay cable lower anchorages at girder level

Shear connectors in form of short pieces of channel steel profiles were welded to the top flanges of both the longitudinal edge beams and the cross beams.

Edge cornices to improve the aerodynamic properties and decrease the instability of the girder section were specified on drawing 039-12-S4A-EST-CHI-51/55 of Nov. 2017. As described in Section 2.1, these cornices were not installed during the erection of the bridge, though.

1.2.3 Stay Cable System

The stay cable system for the Chirajara Bridge consisted of parallel-strand stay cables comprised of multiple seven-wire strands, each protected by a high-density polyethylene (HDPE) sheath. The bundled strands were installed in circular helically-filleted pipe sleeves to improve aerodynamics and reduce vibrations. No compound was present in the void between HDPE-covered strands and the outer sleeves. The number of strands in each cable varied from 4 to 14 strands for the side-span and main-span cables. Anchor cables consisted of 28 strands for each of the 8 cables at both ends. The cables varied in length from 44 m – 153 m.

Active anchors were positioned in the pylon head (see *Fig 1.9*), while passive anchors were placed with eccentricity to the web on the outside of the longitudinal edge beams of the girder (see *Fig 1.7* and *1.8*). Each of the strands in each cable was stressed one by one, using the VSL iso-elongation technique inside the pylon head.

Each stay cable is referenced by a unique letter-number sequence; starting with *"T"* combined with the associated span (A/B/C/D), the consecutive numbering from the stays closest to the pylon (1–15), and the north or south cable respectively (N/S).

During construction, temporary stays cables were installed between the foundation caisson cap and girder, at the point of stays T07B-N/S and T07C-N/S. On both the north and south side, the temporary stays consisted of one cable with 19 strands, 0.6" in diameter and an additional cable of 9 strands, 0.6" in diameter.

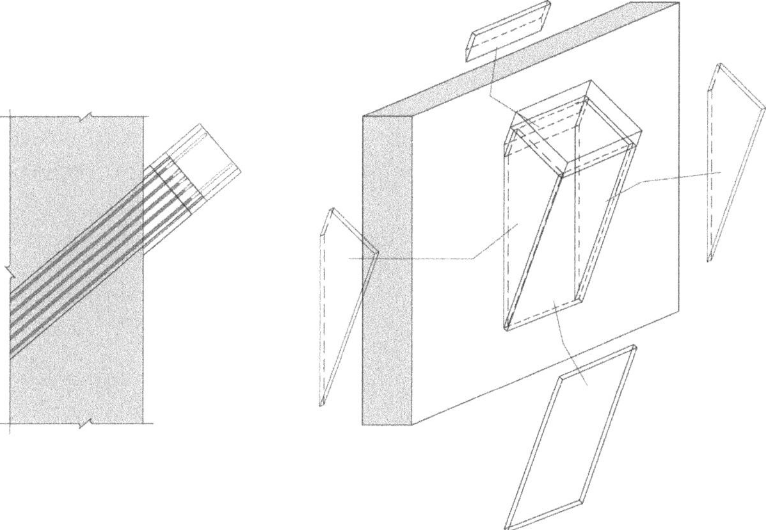

Fig 1.9 Excerpt from design drawings showing the principle of stay cable upper anchorages located within the pylon head

1.2.4 Abutment

Each anchor block was founded on four circular caissons with an outer diameter of 3.0 m and a wall thickness of 0.6 m (see *Fig 1.10*). The outer dimensions of the hollow anchor block were ~12.0 m × 17.2 m × 17.0 m (H × W × L). Apart from the top slab, bottom slab, back wall, and front wall, four side walls ran in the longitudinal direction of the bridge, dividing the anchor block into three hollow chambers (see *Fig 1.11*). All four of these longitudinal walls were prestressed in the vertical direction by three pairs of U-shaped tendons, each consisting of six 0.6" strands.

Within either of the two outermost chambers, two steel I-beams confined within reinforced concrete were placed, referred to as an *anchor beam*. Four stay cables passed through the anchor beam and were anchored on its bottom face. The anchor beam was supported on the top face against the top slab by eight inclined elastomeric neoprene bearings placed in pairs on either side of the anchor cables. The elastomeric bearings were inclined perpendicular to the stay cables at 40.6° with respect to horizontal and placed between two equally inclined steel wedges. One welded to the top of the anchor beam, and the other connected to the bottom face of the top slab.

The anchor beam exited the anchor block through a 1.40 m × 2.17 m hole in the front wall facing the side span and was attached at this end to a concrete transition beam. The transition beam was prestressed in transverse direction to the bridge axis. As the bridge girder and the anchor beams were not constructed at the same level, the purpose of the transition beam was to create a transition between the levels of the girder and the anchor beams.

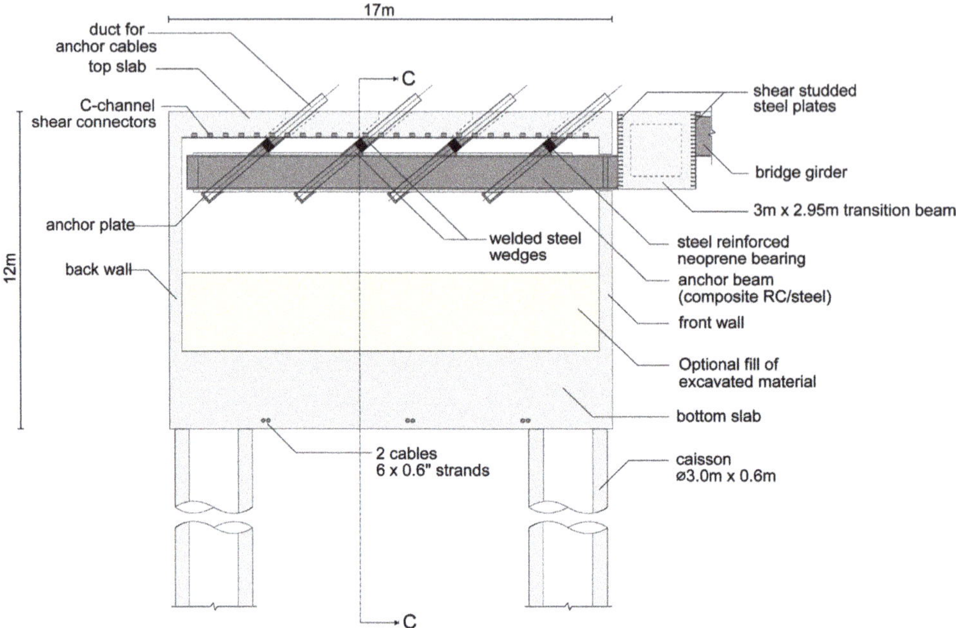

Fig. 1.10   Side view (section B-B) of abutment including anchor block, anchor beam, and transition beam

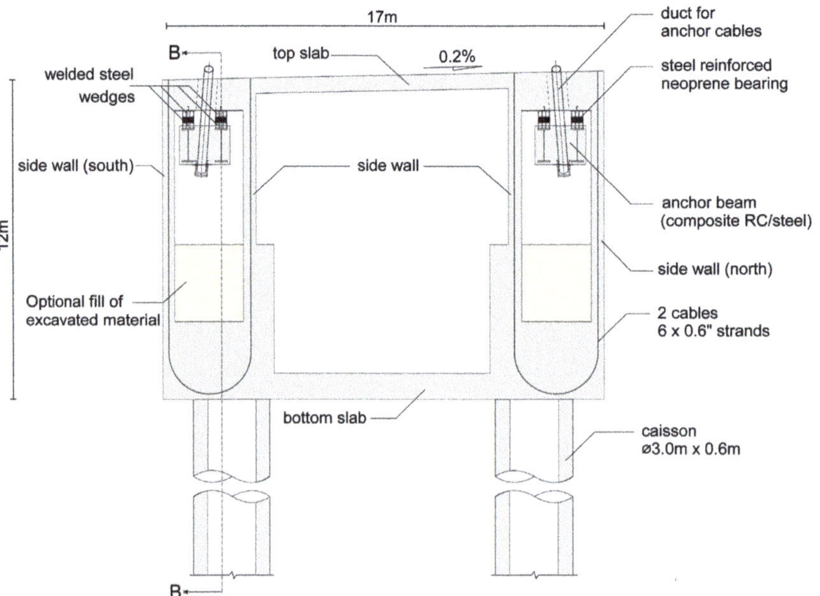

Fig. 1.11   View towards main span (section C-C) of abutment, including anchor block, anchor beam, and transition beam

Chapter 2

Assessment of the Bridge Design

The overall assessment of the bridge design has been performed based on information provided in the design drawings and the engineering design report. The team of international experts commissioned Leonhardt, Andrä und Partner, Germany, to undertake a detailed engineering design check of the bridge, with checks based on *strength load combinations* as required by AASHTO [10]. Additional checks were also made according to Eurocode [8, 9].

2.1 Bearings

An unconventional solution for the bridge bearing design was chosen, containing two elastomeric bearings for vertical support of the girder at each pylon and two inclined elastomeric bearings for each of the stay anchor cables on the anchor beam at the abutment.

Equilibrium of the structure in both longitudinal and lateral direction was achieved through elastic support via deformation of the elastomeric bearings, as no rigid supports, or anti-lifting devices were provided between the girder and the substructure. This resulted in a somewhat *"floating"* statical system leading to excessive displacements for both the service and collapse stages. Generally, the displacements found at the bearings exceed those allowable in both the longitudinal and lateral directions. For service stage loads, it could be expected that some of the bearings would have sheared off.

Expansion, contraction or longitudinal movement of the girder, e.g., due to temperature deviations and breaking forces, would have led to deformation of the elastomeric bearings in shear as well (see *Fig 2.1a* and *2.1b*). Because of the inclination of the bearings, this would have resulted in an upward or a downward displacement of the girder. Hence, these displacements would have led to the girder no longer being level with the rigid abutment. In addition, it was found that lateral (transverse to the bridge axis) displacements of the anchor beam would have been caused by the lateral component of the anchor cable force (see *Fig 2.1c*).

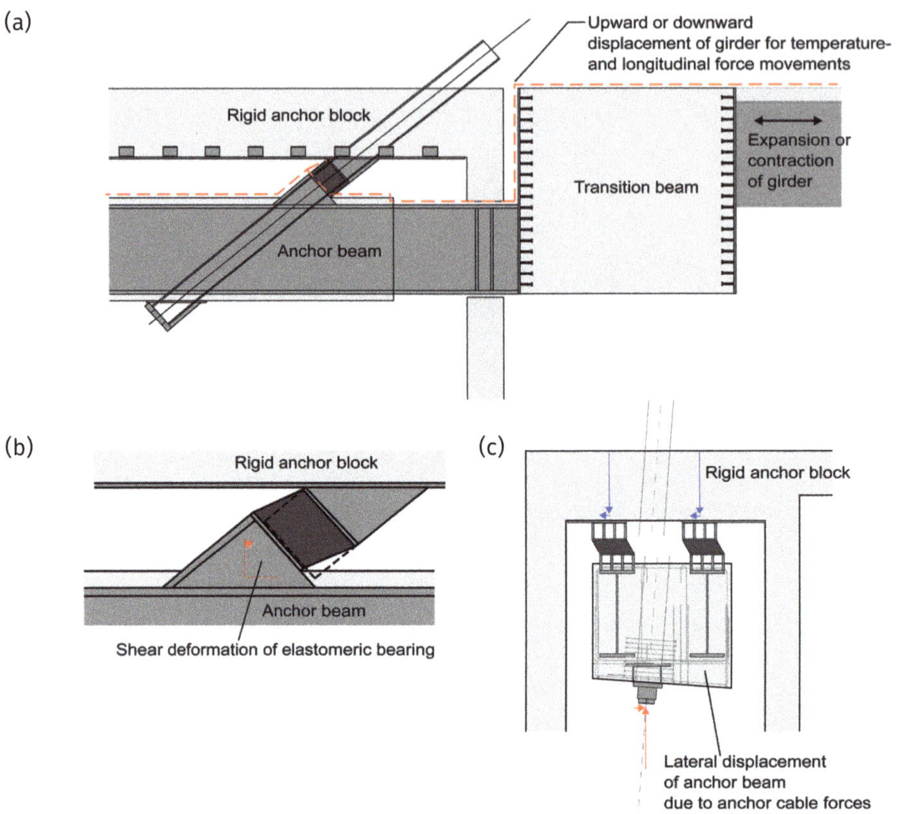

Fig. 2.1   Deformation of elastomeric bearings
        a) Upward/downward displacement due to expansion / contraction of girder
        b) Detailed displacement of elastomeric bearing
        c) Lateral deformation of elastomeric bearings due to lateral anchor cable force

2.2 Abutment

Several deficiencies were found at the abutment. As already stated in Section 2.1, the anchor beam was only supported against the top slab of the anchor block by inclined elastomeric bearings. Besides excessive deformation of these elastomeric bearings, further deficiencies were found regarding the general flow of forces at the abutment (see *Fig 2.2*).

Deficiencies related to the connection between the longitudinal steel beams of the girder and the end plate, connected to the transition beam by shear studs, were identified. In addition to the axial force in the girder, induced by the unbalanced main span, bending moments were identified in the connection between the longitudinal steel beams and the transition beam. The identified bending moments at this point exceeded the bending capacity of the longitudinal edge steel beam of the girder for service stage loads.

Furthermore, no reinforcement went through the connection between the girder and the transition beam. Hence, the entire tension force, introduced from the bending moment at the end of the girder, would have to be transferred through the endplate and the

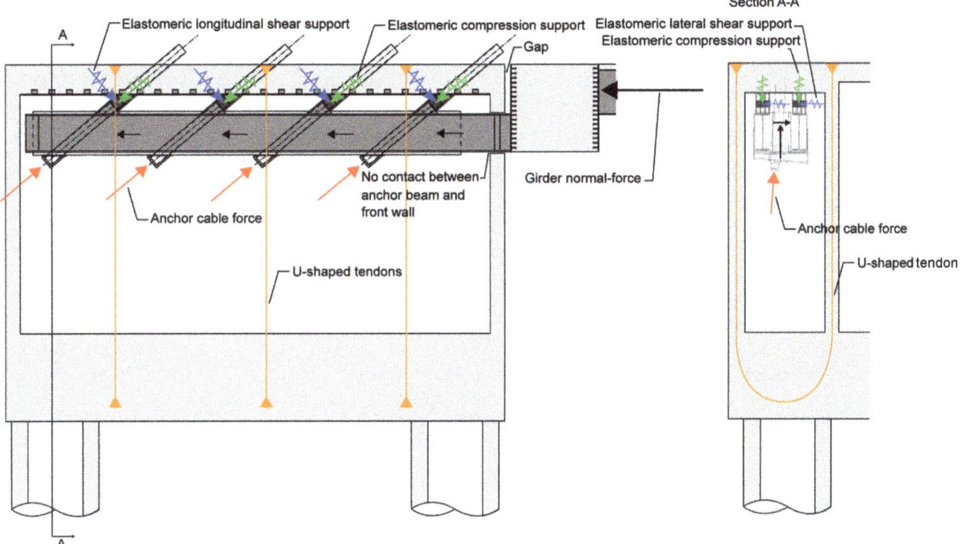

Fig. 2.2 General flow of forces at abutment, for girder axial-force

shear studs into the transition beam. However, the embedment length of the provided shear studs was insufficient to properly transfer these tensile forces, which eventually would have pulled out the studs (see *Fig 2.3*). Also, it was found on site that the shear connectors consisted of bolts with threaded ends, welded to the steel plates (see Section 5.1.5., *Fig 5.30* and *5.32*). Welding of bolts would in general be considered unacceptable for any type of connection, as the tensile capacity of such a connection would be close to zero and the shear capacity undefined.

The transition beam itself was not designed properly for the high bending girder end moments. Deficiencies related to the transition beam were identified by a simple strut and tie model (see *Fig 2.3*). The transition beam was only provided with reinforcement at the surface even though a tension tie would be required at the inner part of the concrete block. This, regardless of positive or negative moment in the connection between the longitudinal steel beam and the transition beam. These deficiencies were confirmed from site visits, where vertical crack formations at the bottom lower corner of the transition beam were identified (see Section 5.2.2, *Fig 5.41*).

The anchor steel beams were not provided with any shear studs. Thus, the axial force, transmitted from the concrete transition beam into the *"composite"* anchor beam, would have to be transferred from the unconfined anchor steel beam to the concrete, purely by friction of bond between steel and concrete. Such a solution would be considered inadequate as pure steel against concrete exhibits rather poor bonding, resulting in a deficient composite effect (see *Fig 2.4*).

Insufficient tensile capacity at the top face of the anchor beam was found. The concentrated compression force, introduced by the anchor cables, must spread out into the two supports at the top of the beam. The provided reinforcement at the top was insufficient to carry the tensile forces, induced by this spreading (see *Fig 2.5*).

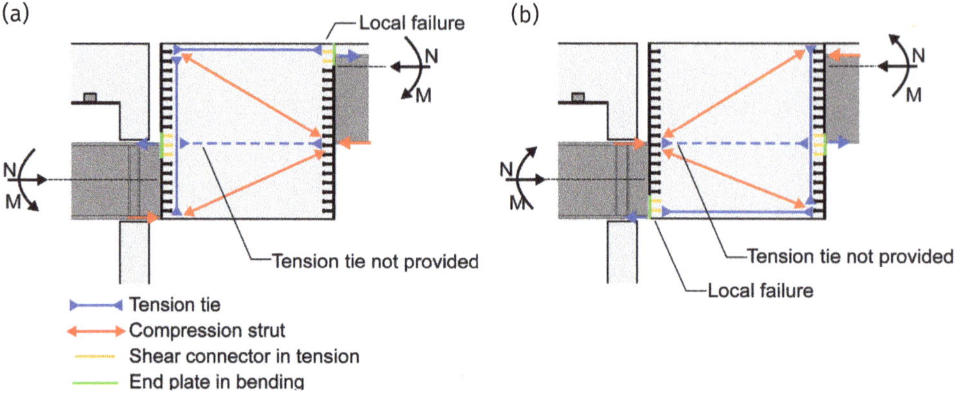

Fig. 2.3 Strut-and-tie model of transition beam
a) Girder exposed to negative moment
b) Girder exposed to positive moment

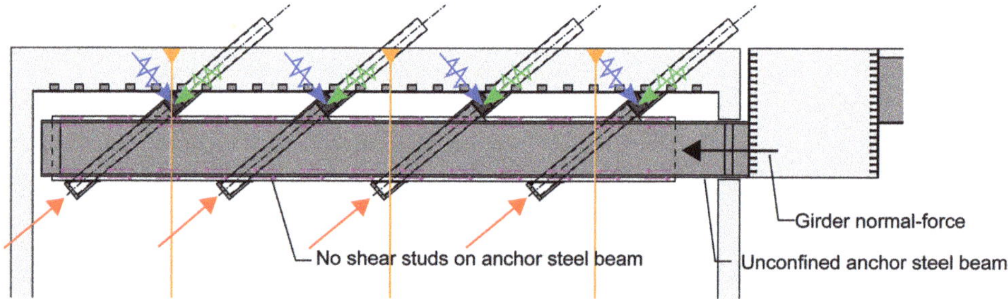

Fig. 2.4 Girder axial force transferred to composite anchor beam purely by bonding as no shear studs were provided

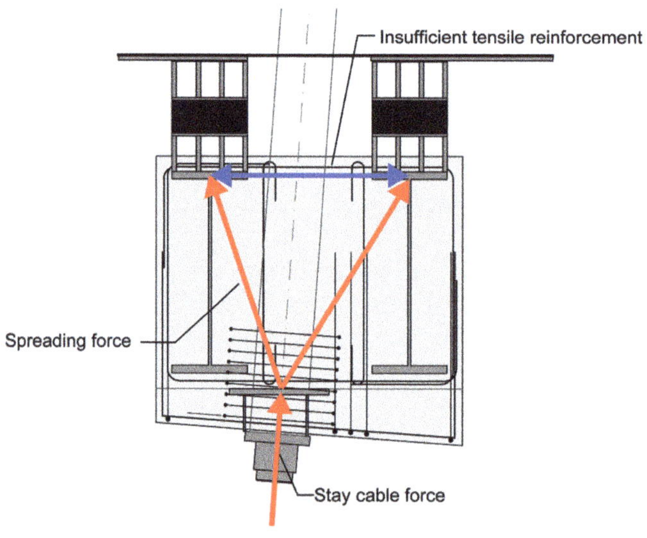

Fig. 2.5 Stay cable anchoring force introducing tensile stresses at top face of the anchor beam due to spreading out compressive forces

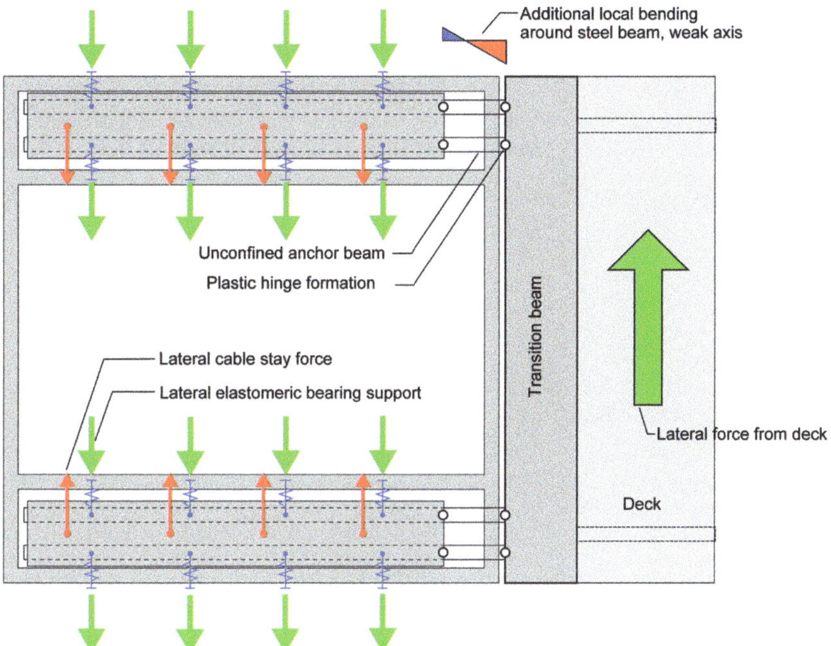

Fig. 2.6 Top view of abutment, highlighting undesirable local bending stresses at naked anchor steel beam for lateral loads on girder

Deficiencies were identified relating to the anchor beam at the abutment. Wind loads and other lateral forces acting on the girder, combined with the horizontal component of inclined stay forces directed inwards, would introduce local bending in the anchor beam. This would have been critical at the connection to the transition beam, where no reinforced concrete encased the I-steel beams of the anchor beam. Edge stresses above the elastic limit were found for service stage loads. This would lead to formation of plastic hinges at the end connecting to the transition beam and eventually also at the end where the anchor beam was covered in concrete (see *Fig 2.6*). The formation of double plastic hinges would lead to instability and failure of the connection due to the column effect at service stage loads.

2.3 Pylon

The overall global capacity of the individual pylon legs has been examined for combined axial force and bi-axial bending throughout their height. Checks for shear capacity according to AASHTO [10] showed insufficient capacity at the base section of the pylon for the service stage design loads, when considering transverse shear parallel to the diaphragm wall.

The local lateral bracing capacity of the link slab and the upper part of the diaphragm were evaluated through a strut-and-tie model (see *Fig 2.7*). The calculation considered a 1:2 distribution of the pylon leg axial stresses into the diaphragm, thus distributed at a

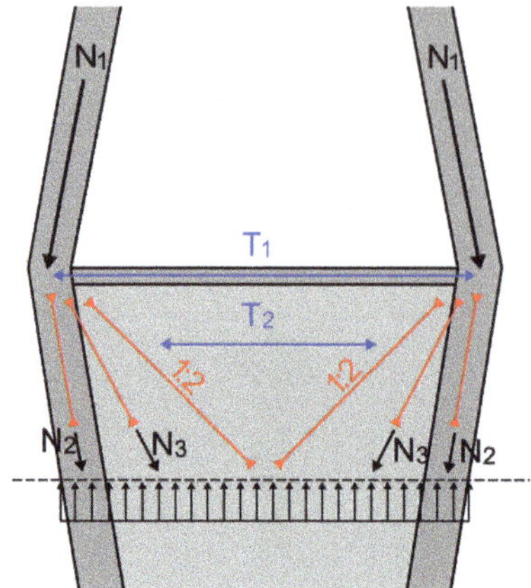

Fig. 2.7 Strut-and-tie model of bracing link slab including tie contribution from diaphragm

depth of ~10 m. Conservatively, the top 3 m of the diaphragm were considered to carry the deviation force in this tie. Insufficient tie capacity was found at the stage immediately prior to collapse, and thus severe lack of capacity was found for service stage design loads. The validity of the strut-and-tie model was confirmed through comparison with the stress distribution found from an elastic FE-model of the legs, diaphragm, and link slab.

It must be noted that these calculations, and the calculations found in the engineering design report, rely on the redistribution of deviation forces into the diaphragm. This redistribution would be necessary as the 12 transverse link slab tendons were found to be clearly insufficient to resist the deviation forces, induced by the directional change of the pylon legs at the knees. This solution is unconventional and differs from the conventional proven solution of carrying all deviations forces directly through knee tie bracing. A further necessity for relying on the diaphragm to carry some of the deviation forces is adequate ductility and deformation capacity of both diaphragm, link slab, and pylon legs respectively. The ability of the structural elements to deform would be an important feature, as the knees would have to move outwards to mobilize these capacity reserves. The engineering design report treats neither the ductility of the pylon legs nor the link slab. In addition, faulty assumptions have been made regarding the ductility of the diaphragm. The faulty assumptions include a linear strain variation from the bottom of the lower pylon legs to the knee and a maximum reinforcement strain of 2.75 % at the knee level (see *Fig 2.8a*). Actually, the strain variation would be non-linear due to the bending of the pylon legs (see *Fig 2.8b*), caused by the combined effect of pylon leg fixity in the caisson cap and the attachment to the diaphragm through the connecting reinforcement. Furthermore, the maximum allowable deformation capacity of the diaphragm would be low due to the very low reinforcement ratio and restraint from the covering concrete and link slab. Thus, the demands for ductility and deformation capacity were not met.

As a consequence of the severe lack of horizontal tie capacity found by the simple strut-and-tie model illustrated in *Fig 2.7*, a detailed non-linear FE analysis, accounting for the actual stress distribution in the combined legs, diaphragm, and link slab, has also been conducted (see Section 9).

Further deficiencies were found when examining the cable stay anchorages within the pylon head. Vertical force components from the stay cables were transferred into the front and back wall only through the shear studs connected to the inside steel box. The horizontal force components from the stay anchorages created bending of the front wall. The solution that was implemented for the anchorages differs from the proven solution of providing transverse steel beams within the pylon head. These would ensure that loads

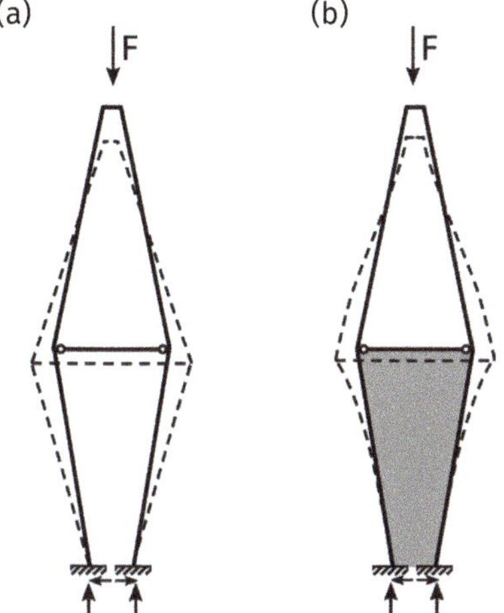

Fig. 2.8 Deformation of pylon with diamond shape configuration for vertical loads
  a) Faulty design assumption, assuming linear deformation of pylon legs and diaphragm
  b) Actual combined non-linear deformation shape of pylon legs and diaphragm

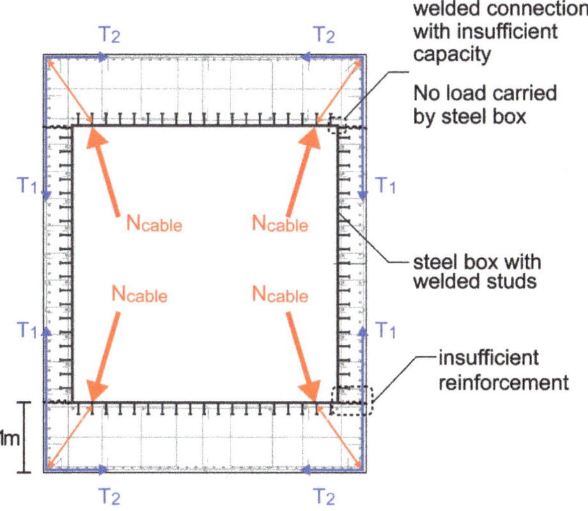

Fig. 2.9 Principal flow of forces due to horizontal components of stay cable forces

are transferred directly into the steel side walls through shear and further to the concrete via shear connectors. A strut-and-tie model revealed that the horizontal force components could not have been transferred safely through the horizontal reinforcement concrete into the pylon head side walls (see *Fig 2.9*). The inside steel box allowed for little to no transfer of stresses. The four steel plates constituting the steel box were only connected by an 8 mm fillet weld in the inside corner in combination with a 3 mm bevel weld on the outside. Both were found to be inadequate.

(a)   (b)

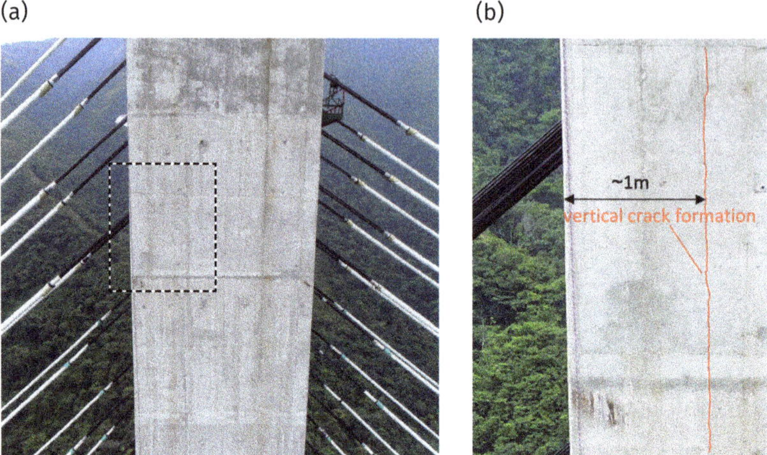

Fig. 2.10   Vertical crack in pylon head
a) North face of pylon head
b) Close-up of north face of Pylon C head with vertical crack formation

(a)   (b)

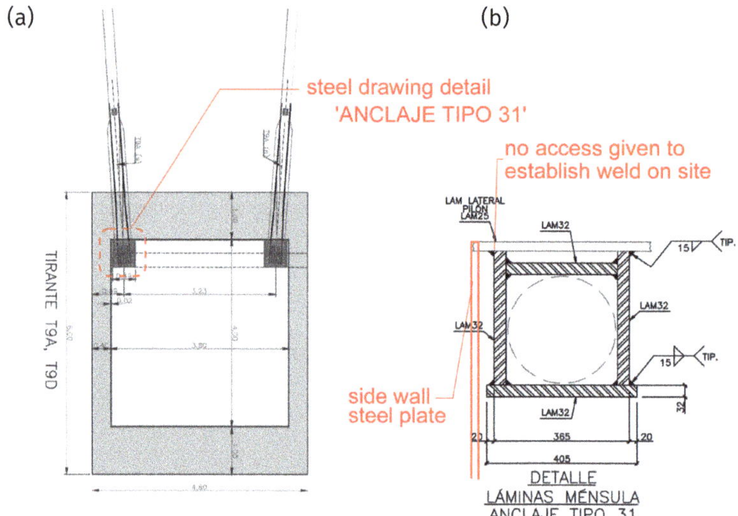

Fig. 2.11   Design drawings showing restricted space for welding of stay anchorage wedges
a) Top view of stay anchorage wedge
b) Top view of steel drawing specifying welds. Addition of present side wall steel
plate shown in red

The faulty design of the pylon head was confirmed by photos taken by drone close to the head of Pylon C. These show the formation of vertical cracking along the inside face of the front and back wall (see *Fig 2.10*).

The steel anchorage wedges were not welded to the pylon head steel box before installation. It is therefore doubtful that the welds next to the side walls could have been properly established due to the restricted space (see *Fig 2.11*).

2.4 Girder

Deficiencies relating to all primary structural components of the bridge girder were found.

Following AASHTO [10], significant exceedance of allowable stresses was found in the bottom flanges of the longitudinal edge beams. Negative bending moments at the pylon supports for service stage design loads would lead to excessive compressive stresses in the bottom flange of the longitudinal edge beams. Additionally, overstress in top flange was found at mid-span for transverse bending moments predominantly, caused by design wind loads at service stage. Shear resistance of the web was found to be insufficient, and the web alone would be prone to plate buckling instability when also checked according to Eurocode 3 [9].

When examining the transverse cross beams, it was found that the elastic capacity of the bottom flange would be exceeded when considering service stage design loads. It should be noted that both the checks of the longitudinal edge beams and the transverse cross beams considered the composite action due to the top concrete slab.

With respect to bolted connections found in the girder, the capacity of the bolted longitudinal edge beam top flange was not found to comply with 75 % capacity of the connecting material as specified by AASHTO [10]. It was also found that the requirement of 75 % capacity was not met by the bottom flange connection of the Type 1 longitudinal edge beam.

Insufficient reinforcement in the concrete slab was identified for the service stage design loads. The required amount of reinforcement has been calculated by superposition of local moments due to the 3 m span between cross beams and global forces and moments found from the FE-model. This superposition identified a lack of longitudinal reinforcement at mid-span, at the abutment, and particularly at the pylons, according to AASHTO [10]. The required reinforcement in the transverse direction due to transfer of shear between the longitudinal steel girder and the concrete slab was also found to be insufficient, according to Eurocode 2 [8].

2.5 Aerodynamic Stability

The girder contained two longitudinal edge beams, connected every 3 m by transverse cross beams of approximately half the height of the edge beams, with a concrete deck on top (see *Fig 2.12a*). Sections between the cross beams comprised only the longitudinal edge beams together with the concrete deck (see *Fig 2.12b*). This type of girder with vertical edge flanges and a thin middle section would be very susceptible to flutter instability compared to other girder types (see type (1), *Fig 2.13*), where positive values of the torsional flutter derivative A_2^* indicate the potential for classic torsional flutter [12].

The eigenfrequencies, found for the service stage indicate a ratio of 1.51 between the first torsional mode and first bending mode. This ratio was found to be 1.36 for the cantilever structure in the stage before collapse. This ratio should be considered critically low for aerodynamic stability. Generally, it would be desirable to attain a ratio greater than 2 between the first torsional mode and the first bending mode [12]. Hence, the risk

(a)

(b)

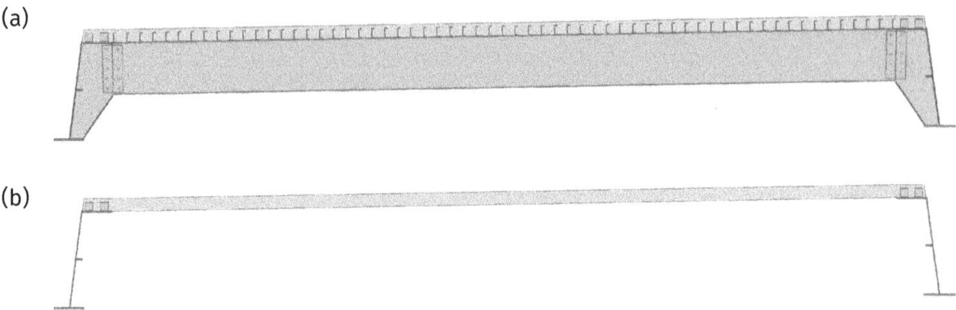

Fig. 2.12 Cross section of girder
a) Section with transverse cross beam
b) Section in between transverse cross beams

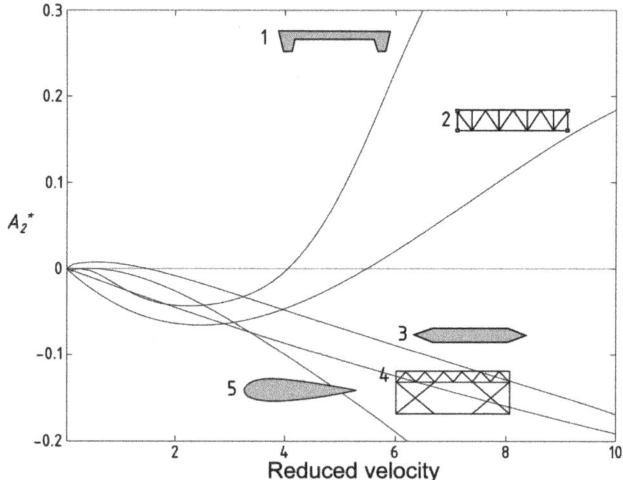

Fig. 2.13 Torsional flutter derivatives for different girder types [12]
(1) unstable bluff box deck
(2) trussed deck with instability
(3) stable streamlined box deck
(4) stable truss deck
(5) thin airfoil

of a coupled flutter occurring would be considered high unless improvements, such as cornices or wind baffles were to be applied to the girder. In addition, it should be noted that four eigenfrequencies were found relatively close to each other, between 0.38 Hz and 0.43 Hz, which would have made the structure very vulnerable to aerodynamic instability.

Wind tunnel tests of the scaled Chirajara Bridge were conducted in 2014 by CSTB, Nantes, France. Results of these tests showed unacceptable aerodynamic behavior of the bridge girder, as originally designed. The original girder section (as during construction) indicated strong vortex shedding excitation. Thus, aerodynamic improvements were suggested to increase stability of the girder. Two vertical wind baffles and edge cornices were recommended by CSTB. However, site visits showed that none of these were im-

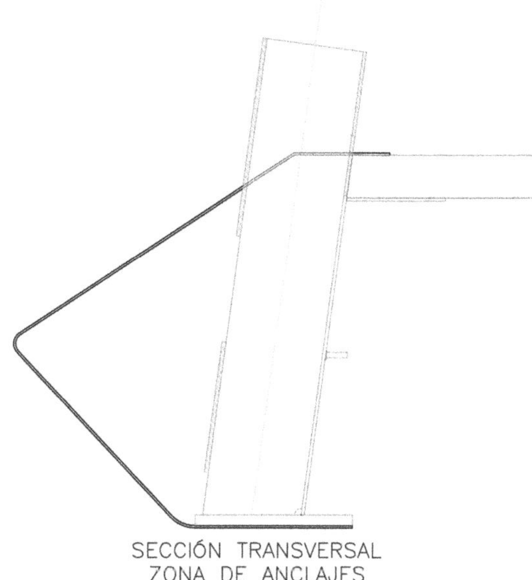

Fig. 2.14 Excerpt from design drawing showing cornices attached on the outside of the longitudinal edge beams

plemented during construction (see Section 5.2.1, *Fig 5.38* and *5.39*). Design drawings indicate the use of cornices for the service stage (see *Fig 2.14*), although no connection details for implementing such edge cornices were found in the steel drawings. Thus, implementation would only have been possible by site welding, which would have been a rather complicated challenge.

Although wind baffles and edge cornices were found to reduce vortex shedding oscillations in most cases, they were found to be relatively ineffective for a -6° wind angle of attack on the girder. Due to the location of the bridge in a steep valley, a -6° wind angle of attack would be considered highly probable.

2.6 Stay Cables

Deficiencies relating to the strength of the stay cables as well as the local anchorages were identified.

For combinations of service stage design loads, exceedance of the maximum admissible tensile force was found for several cable stays, particularly in the main span.

Deficiencies related to the upper stay anchorages at the pylon head are reported in Section 2.1. Regarding the lower anchorages at girder level, the cable stays varied in inclination from ~4–8°, but the longitudinal edge beam had a web with a constant inclination of 7.6° (see *Fig 2.15*). No indication of variation in the inclination of the anchorage boxes was found in the steelwork drawings. This angular difference would lead to bending in the stay cables at the point of attachment. Additionally, the eccentricity of the stay anchorages with respect to the longitudinal edge beam web would create transverse bending in the

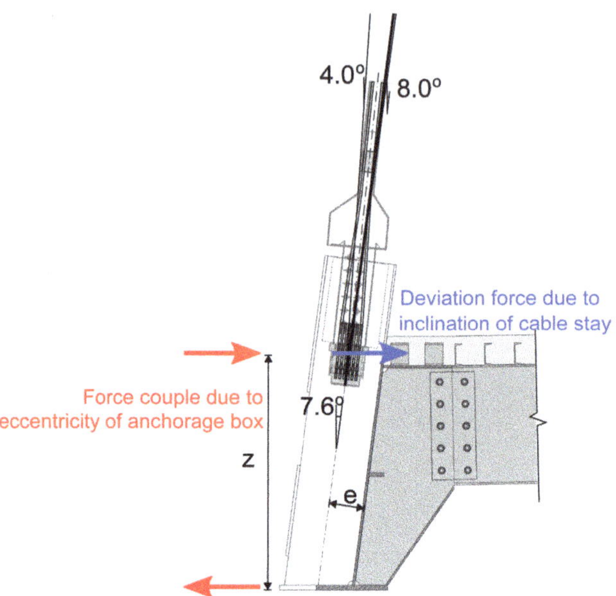

Fig. 2.15 Excerpt of design drawing showing various stay cable inclinations anchored within an anchorage box of constant inclination. Force couple due to eccentricity shown by red arrows

steel works. Particularly, in the bottom flange at points where the anchorage did not coincide with the position of a transverse web beam and accompanying web stiffening plate.

It should be noted that stay forces during construction were not properly documented and apparently provided only through the mobile messaging platform WhatsApp.

2.7 Foundation Caissons

The structural capacity of the caissons was found sufficient during construction and for service loads when examined in bi-axial bending, torsion, and shear. Ignoring the positive effect of the tieback anchors, slight excess of shear capacity for design loads was found.

The installed tieback anchors were not described in the final drawings or the engineering design reports.

2.8 Selected Drawing Inconsistencies

Some contradictions and inconsistencies were detected in the design drawings. Inconsistencies related to the coordinates of the stay cables were found in drawing 039-12-S4A-EST-CHI-24/55. Specified coordinates of the anchor cables T10AN/S and T11AN/S state that these should be anchored outside the abutment (see *Fig 2.16*). In addition, the specifications indicate that all the eastern main-span cables should be anchored at coordinate $Y = \pm6.00$ whereas all other stays are shown anchored at $Y = \pm6.50$.

Inconsistencies between the link slab sections D-D and G-G were found in the drawings. No indication of the location of section G-G was found. Nor was any specified cable curve of the link slab tendons found. In addition, it was found that the link slab tendons transverse to the bridge axis in section D-D should be placed beneath the longitudinal tendons. Section G-G indicates that these transverse tendons should be placed above the longitudinal tendons.

Analysis of caisson foundations including tieback anchors was found in the geotechnical design report. However, no indication of tieback anchors was found in the design drawings of the caissons.

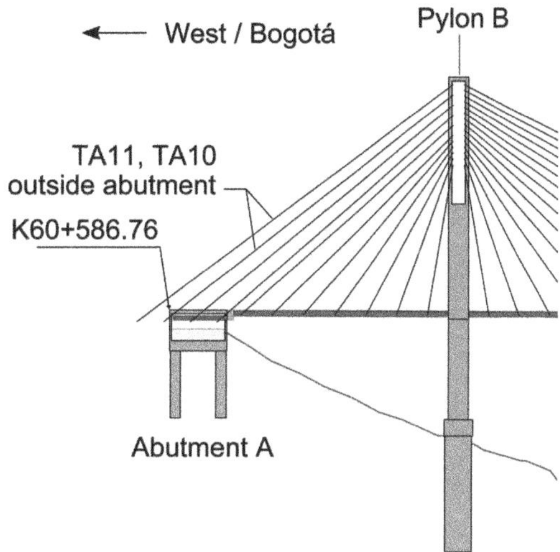

Fig. 2.16 Location of anchor cables according to drawings

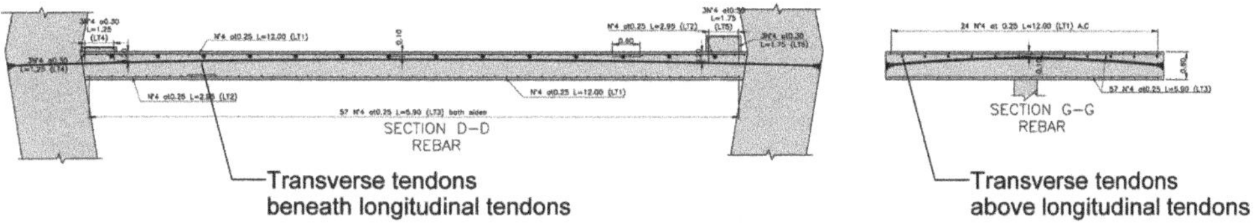

Fig. 2.17 Inconsistency between location of transverse and longitudinal link slab tendons. a) Link slab, transverse section, b) Link slab, longitudinal section.

2.9 General Remarks

Several differences in execution from what was indicated in design drawings were found when comparing the design drawings to observations from site visits and photographic documentation of the construction process.

Inconsistencies were found in the construction of the anchorage block with respect to the drawings. Site observations showed that no nib was cast at the front wall above the exit hole of the anchor beam. Inspection of the inside of both anchor blocks showed that different cable anchorages were constructed on the Bogotá and Villavicencio side respectively (see *Fig 2.18*).

In the main design drawings, regular shear studs are shown between the transverse cross beams and the bridge girder concrete slab. However, in all the steel drawings and, as found on site, channel shear connectors, all aligned in the same direction, were applied (see *Fig 2.19*). This type of shear connector is unconventional but covered by AASHTO [12].

Drawings of the pylon caissons show no specification of tieback anchors, which were documented by site visits and photographic records of the construction process (see *Fig 2.20*).

Villavicencio / East side
Anchorage not according to drawings

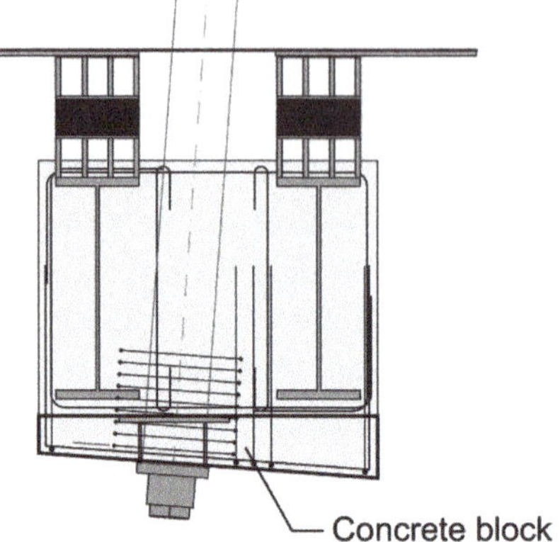

Bogotá / West side
Anchorage according to drawings

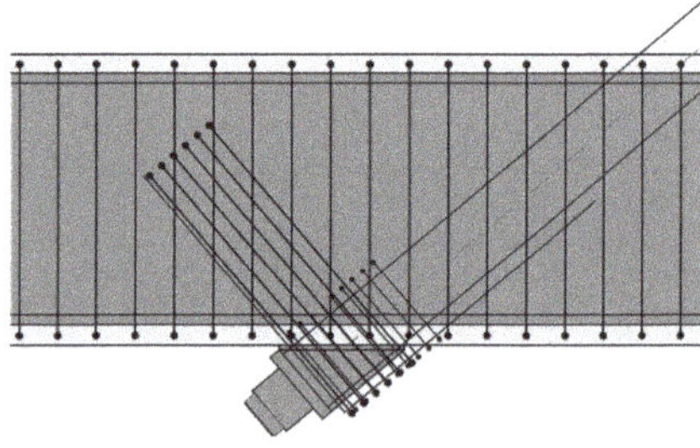

Fig. 2.18 Photos of stay cable anchorages inside both anchor blocks and excerpt from design drawings

Fig. 2.19 Photo of channel shear connectors welded to transverse cross beams

Drawings also specify temporary bracing during construction by strands connecting the top of the caisson cap to the girder at the location of cable stays T07B-N/S and T07C-N/S. Photos taken during the construction process show that the total number of specified strands was probably not provided and that several of the strands were not in tension, i.e., slacking (see *Fig 2.21*). Documentation on the tensioning scheme and when the strands were tensioned was not provided.

Fig. 2.20 Photo of inside of pylon caisson during construction, showing protruding strands from installed tieback anchors

(a)

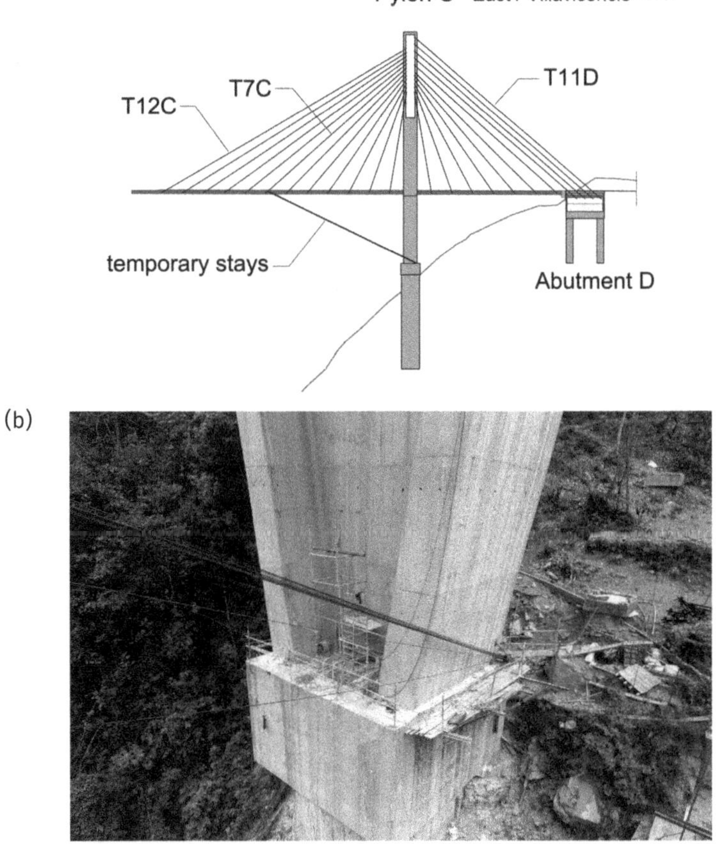

(b)

Fig. 2.21 Temporary stays
a) Location of temporary stays
b) Photo showing tensioned and slacking strands of temporary stays of Pylon C

2.10 Summary of Design Deficiencies

2.10.1 Critical Deficiencies

1. A severe lack of bracing tie capacity of the **link slab and diaphragm** was found for design loads, service loads, and during erection. The lack of bracing was the main deficiency leading to the eventual collapse of Pylon B and the bridge as a whole.

2. The bearing layout at pylons and abutments would result in a *"floating"* statical system. An unproven solution that amongst other things would induce large stresses in parts of the structure with incomplete direction and without respect for the flow of forces in the complete system. The **bearings** would not have sufficient capacity to withstand the expected shear forces and deformations during construction and for the service stage design loads, resulting in excessive shear deformations with potential failure.

3. Considerable redesign and retrofit of the **anchor beam** assembly would be needed to accommodate the design loads in the service stage. Deficiencies were found regarding the formation of local plastic hinges resulting in instability, insufficient lateral reinforcement at top face, and missing shear connectors between steel I-beams and encasing concrete to provide composite action. Most importantly, the bearing design related to the transition beam would need a complete redesign.

4. Longitudinal force components would create bending and high local shear stresses in the front and back wall of the **pylon head** since no bracing with sufficient shear capacity was provided inside the pylon head to transfer horizontal and vertical components of stay forces to the pylon head concrete walls.

5. The **transition beam** and corresponding connections were found to be inadequate and not designed with proper tensile capacity.

6. As designed, the **longitudinal edge beams of the girder** would experience a significant exceedance of allowable stresses in the bottom flange, due to negative bending at the pylon supports for service stage design loads. Overstress in the top flange was found at mid-span for transverse bending moments caused by design wind loads at service stage. Shear resistance of the web was found to be insufficient, and the web would be prone to buckling instability, even with the provided stiffening.

7. A lack of reinforcement was found in the longitudinal direction for the **girder concrete slab**, in regions with large negative moments. The specified transverse reinforcement was insufficient when considering transfer of shear between steel girder and concrete slab.

2.10.2 Secondary Deficiencies

1. Sufficient structural capacity during construction and for service loads was found for the **caisson**. A slight excess of shear capacity was found for design loads. This could be rectified by applying actual material strengths instead of considering design values

2. The elastic capacity of the **girder cross beams** would have been exceeded for service stage design loads.

3. **Bolted connections in the girder** would not comply with 75 %-capacity of connecting materials according to AASHTO.

4. The girder section as erected was considered **aerodynamically unstable** and would be very prone to vortex shedding excitation. Proposed baffles and cornices were not installed during erection.

5. The varying transverse inclination of **stay cables** combined with the constant inclination of anchorage boxes at girder level could lead to bending of the cables at the point of attachment and/or local deformation of the steel beams.

Chapter 3

Status of Construction and Loads Immediately Prior to Collapse

3.1 Construction Sequence and Status Immediately Prior to Collapse

The general construction sequence of the main structural components is presented in *Fig 3.1*. The construction status at the end of specifically chosen months is presented in *Fig 3.2* to show how the structure was progressing during construction.

As the side spans (AB and CD) were erected before cable installation, the girder was not erected by the typical balanced cantilever method. Although this method was initially proposed, the side spans were eventually erected on temporary supports to save time (see *Fig 3.2c and 3.3a*).

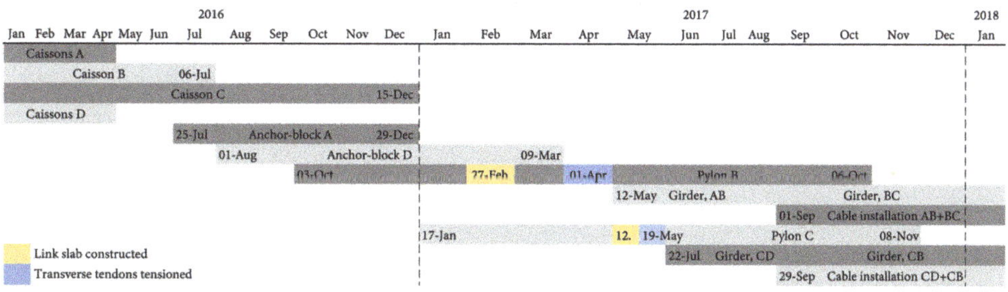

Fig. 3.1 Overview of the general construction sequence

As seen in *Fig 3.3b*, the main-span sections were erected by use of trolley cars. These were supported by cables, anchored in the anchor blocks, and supported at the diamond top of the pylon legs. To balance the cantilever main-span sections, the stay cables were tensioned immediately after pouring of concrete on each of the girder sections. The temporary supports at the side-span sections were released immediately before tensioning of the respective side-span stay cables (see *Fig 3.2c*).

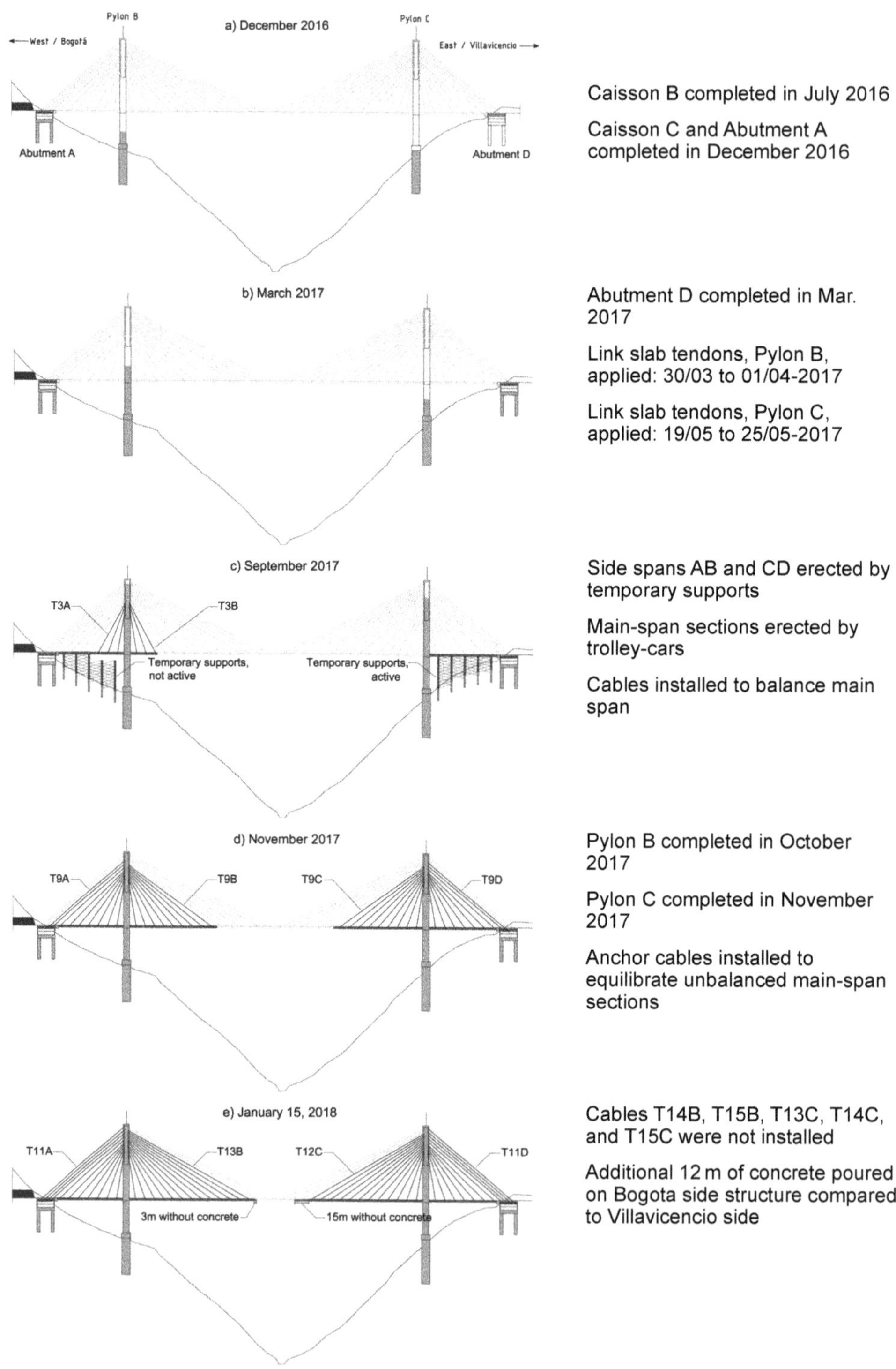

Fig. 3.2   Illustration of the construction status at the end of specific months

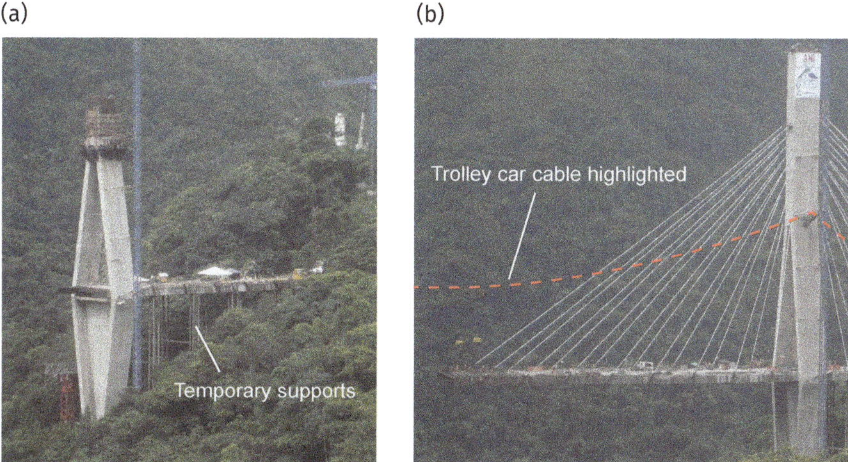

Fig. 3.3 Erection of girder
  a) Erection of side span by temporary supports
  b) Erection of main span by trolley car

Fig. 3.4 Status of construction immediately prior to collapse
  a) Full bridge
  b) Zoomed view of main span end, east structure
  c) Zoomed view of main span end, west structure

Beyond the structural deadload of the pylons themselves, the vertical forces introduced to the pylons were slowly increasing along with erection of main-span sections and installation of stay cables.

The status of construction on January 15, 2018, immediately prior to collapse can be seen in *Fig 3.2e* and *3.4*, where the latter comprises photo extractions from the safety camera video, recorded on the day of the collapse. The two structures were almost at the same stage just prior to the collapse. Stay cables T13B(n,s) were installed and a segment of 12 m concrete was poured on the Bogota side structure two days before the collapse. As can be seen from *Fig 3.4b*, this work was currently going on at the Villavicencio side structure, which would have left the two structures at the same stage if the work was not interrupted by the collapse an hour later. However, the stay cables T13C(n,s) were not yet installed and the 12 m of concrete was not yet poured on the Villavicencio side structure at the time of collapse.

3.2 Loads Immediately Prior to Collapse

3.2.1 Structural Deadload

The estimated structural deadload of relevant structural elements, present immediately before collapse, is summarized for the Bogota and Villavicencio side structure in *Tables 3.1 to 3.5*.

Tab. 3.1 Structural deadload of girder immediately prior to collapse, Bogota side

Girder, Bogota side	Length [m]	Width [m]	Thickness [m]	Area [m²]	Volume [m³]	Weight [kN/m³]	[kN/m²]	[kN/m]	[kN]
Long. steel beams	387.0			0.062	24.1	78.0		10	1878
Trans. steel beams / 3 m	845.0			0.015	12.9	78.0		5	1009
Concrete	190.5	13.0	0.2	2477	495.3	25.0		65	12383
							6	80	15269

Tab. 3.2 Structural deadload of girder immediately prior to collapse, Villavicencio side

Girder, Villavicencio side	Length [m]	Width [m]	Thickness [m]	Area [m²]	Volume [m³]	Weight [kN/m³]	[kN/m²]	[kN/m]	[kN]
Long. steel beams	387.0			0.062	24.1	78.0		10	1878
Trans. steel beams /3 m	845.0			0.015	12.9	78.0		5	1009
Concrete	178.5	13.0	0.2	2321	464.1	25.0		65	11603
						Sum	6	80	14489

Tab. 3.3 Structural deadload of stay cables immediately prior to collapse, Bogota side

Stay Cables, 0.6", Bogota side	Length [m strand]	Weight [kN/m strand]	[kN]
Anchor Cables, T8A(n,s)–T11A(n,s)	23610	0.011	255
Cables, T1A(n,s)–T7A(n,s)	7488	0.011	81
Cables, T1B(n,s)–T13B(n,s)	24788	0.011	268
		Sum	605

Tab. 3.4 Structural deadload of stay cables immediately prior to collapse, Villavicencio side

Stay Cables, 0.6", Villavicencio side	Length [m strand]	Weight [kN/m strand]	[kN]
Anchor Cables, T8D(n,s)–T11D(n,s)	23610	0.011	255
Cables, T1D(n,s)–T7D(n,s)	7488	0.011	81
Cables, T1C(n,s)–T12C(n,s)	21006	0.011	227
		Sum	564

Tab. 3.5 Structural deadload of Pylon B identical to Pylon C immediately prior to collapse

Pylon	Height [m]	Width [m]	Thickness [m]	Area [m²]	Volume [m³]	Weight [kN/m³]	[kN]
Solid top piece	1.0	4.6	6.0	27.6	27.6	25	690
Hollow mast	34.9	4.6	6.0	12.4	432.1	25	10804
Hollow bottom piece	2.2	4.6	6.0	22.0	47.3	25	1183
Top leg	34.3	1.6	6.0	19.2	658.2	25	16454
Bottom legs	30.1	1.6	6.0	19.2	577.2	25	14429
Diaphragm	29.5	14.2	0.5	4.7	139.3	25	3482
Slab	0.6	14.2	6.0	84.9	50.9	25	1274
						Sum	48315

From *Tables 3.1* to *3.5*, it can be seen that additional 821 kN of structural deadload was present at the Bogotá side structure compared to the Villavicencio side structure immediately prior to collapse.

3.2.2 Construction Live-load

No unusual work was going on at the time of collapse, which was also verified by the safety camera on site. A relatively minor construction live-load of in total 400 kN was acting on the Bogota side structure immediately prior to collapse. This value has been based on an excel document, describing the actual construction live-loads immediately prior to collapse, provided by GISAICO.

3.2.3 Wind and Temperature Conditions

Measurements were taken on site to monitor the wind and temperature conditions during construction. The available data contained measurements from the top of Pylon B of the mean wind, wind gust, and temperature over a period of 30 min. The wind measurement data are presented in *Fig 3.5* and *3.6* for the day of the collapse and a few days before the collapse respectively.

No unusual wind conditions were found by the measurements from the day of collapse, which indicates average wind speeds between 1 m/s and 3 m/s. Nor anything unusual was found from the measurements in the days before the collapse.

The temperature measurements prior to the collapse neither indicate any unusual values nor extreme changes (see *Fig 3.7*).

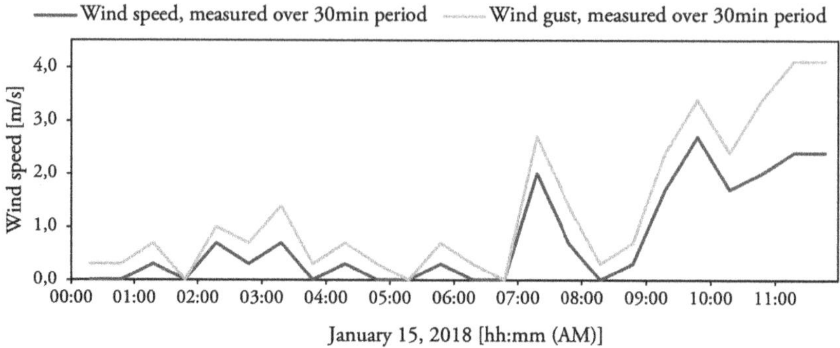

Fig. 3.5 Wind measurements over 30 min period on top of Pylon B on the day of collapse

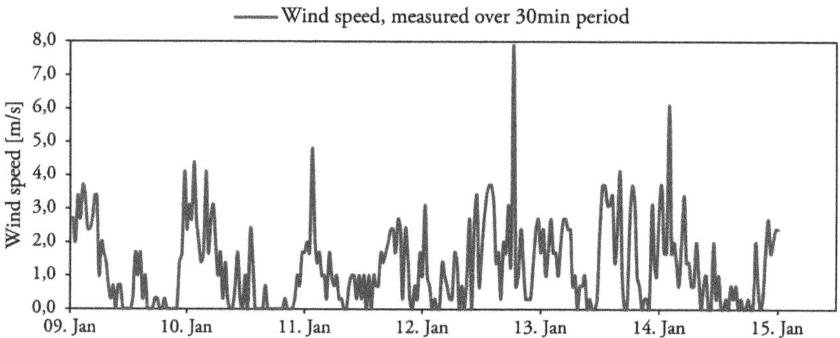

Fig. 3.6 Wind measurements over 30 min period on top of Pylon B, few days prior collapse

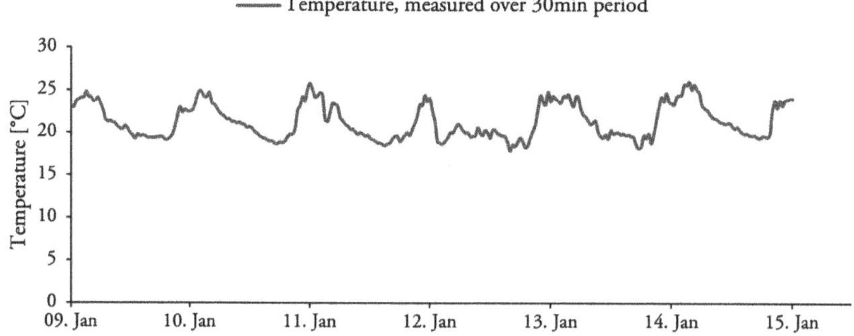

Fig. 3.7 Temperature measurements over 30 min period, a few days prior to collapse

3.2.4 Seismic Records

According to the Colombian Geological Service (Servicio Geológico Colombiano), a total of 43 minor seismic events were recorded in Colombia on the day of the collapse (January 15, 2018). The recordings range from 1.1 to 3.4 in local magnitude (Richter), M_L.

Three minor seismic events were recorded on the day of the collapse within a radius of 200 km from the Chirajara Bridge, the closest being located 118 km from the location of

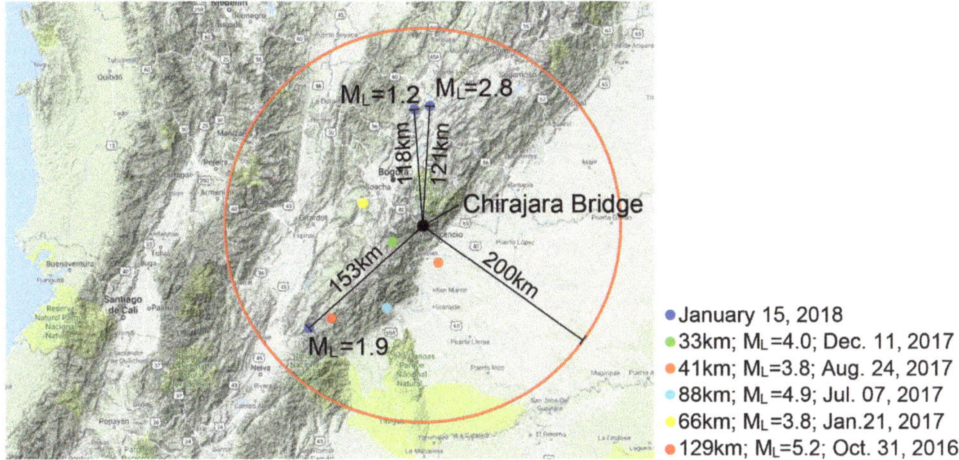

Fig. 3.8 Relevant seismic recordings in a radius of 200 km from Chirajara Bridge

the bridge (see *Fig 3.8*). Due to the low magnitude and the large distance from the location of the bridge, it was concluded that these seismic events did not influence the structure.

Considering the construction period from August 2015 to January 15, 2018, a total of 32 events, with a minimum local magnitude of 3.5, have been recorded within a radius of 200 km from the location of the bridge. Relevant recordings from this period are shown in *Fig 3.8*. The maximum local magnitude (M_L=5.2), recorded in this period took place on October 31, 2016, at a distance of 129 km from the location of the bridge. Four seismic events closer to the bridge (<100 km) were also recorded in this period. However, all earthquakes recorded during the construction process were of relatively low magnitude and not in the immediate vicinity of the bridge. Thus, it was found that these seismic events could not have influenced the bridge or the foundations.

Chapter 4

Description of the Collapse

The nature of the collapse of Pylon B can be established in two ways. One is by assessing the footage from a nearby traffic camera, which had the majority of Pylon B and the western main- and back-span in its field of view and filmed during the entire collapse. The second is by evaluating the orientation and placement of the various structural elements from the debris zone after the collapse.

4.1 Video of Collapse

The team was provided with video footage of the collapse and 55 minutes leading up to the collapse [D10]. The video camera was situated next to a nearby road north of the bridge filming towards south (see *Fig 4.1* and *4.2*). The video was filmed at ~15 frames per second with a resolution of 1280×720 pixels. The part of the view containing the bridge has a resolution of 285×285 pixels. The video was split into separate frames and then stabilized and corrected by image translation and rotation with the software Hugin to reduce wind-induced camera motions.

Fig. 4.1 Still frame from traffic camera video immediately before collapse

Fig. 4.2   Photo of traffic camera mast (right)

The following general conclusions can be made based on observations from the video:

- In the hour leading up to the collapse, the movement and swaying of trees indicate wind speeds corresponding to gentle to moderate breeze. The video shows no indication of strong winds or storm conditions.
- Immediately following the collapse, no oscillating vibrations of the traffic camera mast can be observed which could otherwise have indicated a sudden ground movement.
- The duration of the collapse is approximately 7 seconds, from the first indication of collapse initiation, i.e., slacking of the shortest main-span stay cables, to the complete collapse by the pylon head hitting the ground.

In the following some selected still frames are described and assessed. The interval between each of the chosen frames varies from 333 to 933 milliseconds, as seen in the middle column. Note that the right-hand image of each row is copied into the left-hand column of the next row. This allows for both vertical and horizontal reference comparisons between frames.

The collapse initiates on Jan. 15, 2018, at 11:48:50 am. This is set as zero-time reference. First visible sign of collapse initiation is slacking of two shortest southern main-span cables.

0.000 s   →   0.533 s   →   0.533 s

0.533 s → 0.600 s → 1.133 s

Slight slacking of two shortest southern side-span cables.

Early separation between leg and diaphragm visible.

1.133 s → 0.933 s → 2.067 s

Slacking of all southern side-span cables except anchor cables.

Downwards inclination of main-span girder. Side-span girder remains stationary.

Diamond shape expanding. Minor shift of pylon head downwards and towards east.

2.067 s → 0.733 s → 2.800 s

Increasing drop of main-span girder.

Visible signs of southern leg having completely detached from diaphragm along whole height.

Few anchor cables start slacking.

Significant vertical and eastern movement of pylon head.

2.800 s → 0.533 s → 3.333 s

Signs of hinge formation in northern knee.

Severe kink of girder at pylon axis.

Continuation of downwards east pylon head movement.

Diaphragm and side-span girder remain stationary.

3.333 s → 0.467 s → 3.800 s

Only the two longest anchor cables remain in tension.

Angle at northern knee increases significantly.

First signs of dust below main-span girder, possibly from contact with the link slab.

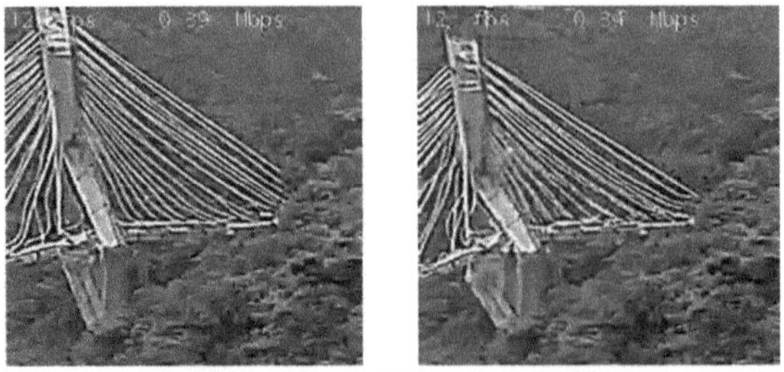

3.800 s　→　0.600 s　→　4.400 s

Pylon head no longer shifts towards east but drops completely vertically.

The girder is no longer continuous, having separated at the pylon axis.

Single anchor cable in tension.

Hinge formation between upper northern pylon leg and pylon head.

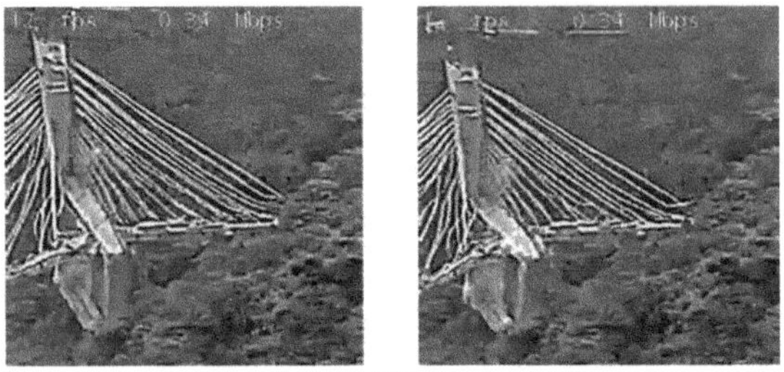

4.400 s　→　0.333 s　→　4.733 s

Upper northern pylon leg starts to collide with the girder above the link slab.

All anchor cables slack indicating a free fall of the pylon head.

Side-span girder remains level. The diaphragm is partly hidden by dust but considered to still be upright.

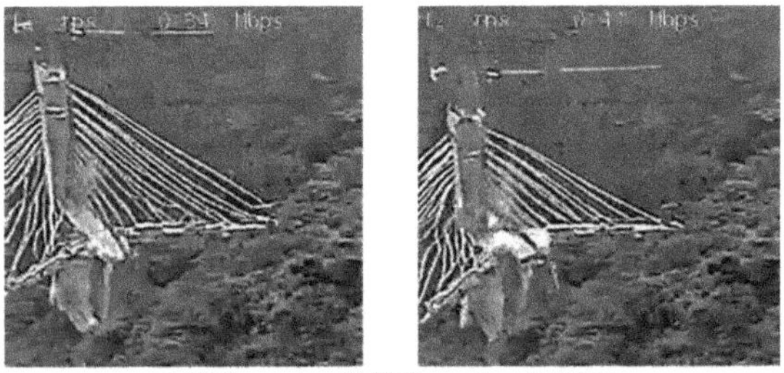

4.733 s　→　0.533 s　→　5.267 s

The pylon head collides with the link slab.

First visible inclination of the side-span girder, indicating loss of support at the diaphragm / pylon axis.

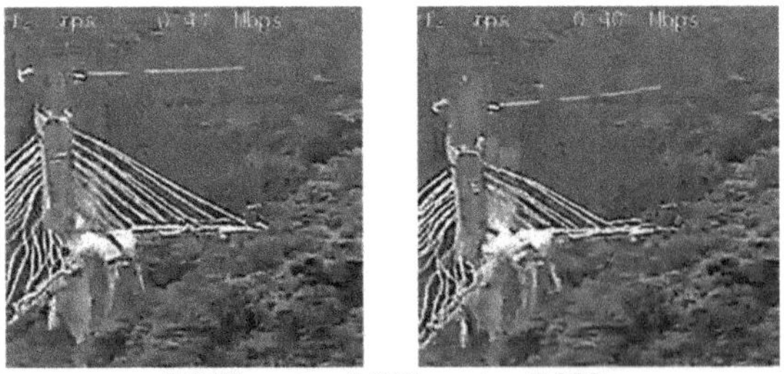

5.267 s　→　0.333 s　→　5.600 s

Pylon head continues its free fall with its bottom beginning to impinge on the hanging main-span girder.

Further drop of the side-span girder which is seemingly still supported at the abutment.

Pylon legs and diaphragm is mostly hidden by dust.

5.600 s　→　0.733 s　→　6.333 s

The bottom of the pylon head meets ground level.

Diaphragm and northern leg are hidden behind concrete dust.

The side-span girder continues its decline at the pylon axis.

Total collapse of Pylon B. Neither cables, main-span girder, side-span girder, or any part of the pylon are visible due to dust

6.333 s → 0.600 s → 6.933 s

4.2 Orientation and Placement of Debris

From the orientation and placement of structural components constituting the pylon, it is evident that the pylon experienced a nearly vertical drop. The girder likewise came down vertically and ended up in the debris zone along the original longitudinal axis of the bridge.

The pylon head, pylon upper legs, pylon lower legs, link slab, and diaphragm all came to rest within a zone approximately 70 m × 37 m east to the caisson cap. The lengths were established based on the known width of the pylon head as reference (see *Fig 4.3*). It is evident how the pylon diamond shape expanded during collapse, as the pylon head is situated directly between the southern and northern pylon legs. The centroid of the pylon head is located just 20 m east of the center of the caisson cap. Slight sliding to the east could have occurred due to the steep ~33 % grade slope of the terrain surrounding the caisson.

The only discrepancy from a completely vertical drop is the upper northern leg being centered rather than situated north of the caisson cap and the diaphragm with link slab looking as if it tilted towards north-east rather than straight along the longitudinal axis of the bridge.

Fig. 4.3   Top view. Outline of zone containing pylon debris. Measurements based on tower width reference (red)

Fig. 4.4   Photo of Pylon B taken a month prior to collapse including indication of structural elements

Apart from the southern pylon leg, which is severely damaged, most of the structural elements seem to have failed primarily in the construction joints following either hinge formation or shearing of the joint. Thus, the diaphragm has been separated along the outer edges where it was connected to the lower pylon legs and in the bottom from the caisson cap. The pylon legs have detached in the lower and upper ends from the caisson cap and the pylon head respectively and separated in the knee joint between upper and lower legs. The post-tensioned link slab has not been separated from the diaphragm but has been separated in the knee joints on both sides.

An overview of the debris zone with an indication of the various structural elements and a reference to the same elements in the state prior to collapse is provided in *Fig 4.4* and *4.5*.

Fig. 4.5 Top view of Pylon B debris zone including indication of structural elements (taken Jan. 17, 2018)

Chapter 5

Site Visits

Collaborators of ONC were on site on several occasions, including the days following the collapse, on January 16–17 and on March 1. Photographic records and drone videos from these visits were provided to the team.

Members of the team were allowed access to the site on two separate occasions, namely March 20 and March 22. The first visit included access to both the standing Pylon C and Abutment D as well as the collapsed Pylon B and Abutment A. The second visit included access to some of the more difficultly accessible elements in the debris zone.

5.1 Pylon B and Abutment A

5.1.1 Pylon Foundation Caisson

The foundation caisson cap and the part of the circular mono-caisson extending above ground showed no sign of damage, settlements, or movements. Only superficial damage to the caisson cap was detected (see *Fig 5.1* to *5.3*). Severe damage was found on top of the caisson cap in the construction joint with the lower pylon legs and the diaphragm (see *Fig 5.4*). Damage in the construction joint included crushing of concrete and rupturing of connecting reinforcement.

A brief internal inspection of the circular caisson revealed no sign of damage or variations on the inside caisson wall.

5.1.2 Lower Pylon Legs and Diaphragm

It was observed on site that both the southern and northern lower pylon legs had separated from the diaphragm along the entire length of the legs rupturing all connecting horizontal reinforcement immediately at the inner face of the lower pylon leg (see *Fig 5.5* to *5.9*). No reinforcement pull-out failure was observed. Apart from the separation from the diaphragm, both lower pylon legs showed severe damage by concrete crushing and reinforcement rupture at the base and knee level, resulting from hinge formation at these joints (see *Fig 5.10*, *5.12*, and *5.13*). Additionally, the southern lower pylon leg showed severe damage by concrete crushing and reinforcement rupture, localized around mid-height of the leg (see *Fig 5.11*). It is possible that this damage was a result of the leg impacting onto the adjacent crane foundation during collapse. Furthermore, local concrete spalling was observed at the edges of either lower leg.

It was evident from photos taken on site that the diaphragm reinforcement ratio was low as the concrete area was relatively large compared to the protruding reinforcement (see *Fig 5.7*).

A thorough mapping of horizontal reinforcement bars connecting either of the lower pylon legs with the diaphragm was carried out at the request of the team. For the critical uppermost 6 m of the connection, the mapping showed an average rebar spacing of 195 mm between the southern lower leg and diaphragm, while an average spacing of 207 mm was found between the northern lower leg and diaphragm. This should be compared to a uniform spacing of 200 mm, indicated on the design drawings. The variation showed a minimum and maximum spacing of 70 mm and 280 mm respectively for the southern connection, while a minimum and maximum spacing of 180 mm and 350 mm respectively was found for the northern connection.

A thorough mapping of crack formation in the upper part of the diaphragm was carried out at the request of the team (see *Fig 5.14*). This mapping showed no distributed vertical crack pattern, as could be expected from a ductile diaphragm carrying increasing lateral deviation forces. Instead, the mapping revealed a diagonal crack pattern, with cracks of minor width, originating in the upper northern corner. Some vertical cracks were detected in the upper southern corner of the diaphragm. Based on the lack of vertical crack formation in the upper northern corner of the diaphragm and the absence of distributed vertical cracks observed on the diaphragm of Pylon C (see *Fig 5.49* to *5.52*), it is believed that these vertical cracks resulted from either the impact of the pylon head with the southeast part of the link slab during collapse or the diaphragm impacting with the ground at the end of collapse.

Fig. 5.1   Caisson cap south face with no apparent damage

Fig. 5.2 Superficial damage detected on the lower east face of the caisson cap

Fig. 5.3 No signs of damage in construction joint between circular caisson and caisson cap

Fig. 5.4 Top southern part of caisson cap. No signs of damage to the caisson cap. Severe damage by concrete crushing and reinforcement rupture, where the northern lower pylon leg was connected. Longitudinal reinforcement of lower leg bent downwards. Anchorage of temporary stays visible (14 strands detected)

Fig. 5.5 Aerial photo of the diaphragm and the northern lower pylon leg. Visible separation of the diaphragm from both the southern and northern lower legs

Fig. 5.6 Photo of separation between diaphragm and northern lower pylon leg. Horizontal connecting reinforcement ruptured along the entire lower leg

Fig. 5.7 Diaphragm. Separation of the diaphragm and the northern lower pylon leg

Fig. 5.8 Southern lower pylon leg. Separation from the diaphragm along the entire leg. Spalling of the concrete along the edges of the lower leg

Fig. 5.9 Lower southern pylon leg. Yielding and necking of reinforcement connecting the diaphragm and the lower southern pylon leg

Fig. 5.10 Base of the southern lower leg

Fig. 5.11 Damage apparent on the southern lower leg at mid-height

Fig. 5.12   Southern lower leg at knee level

Fig. 5.13   Northern leg just above base level

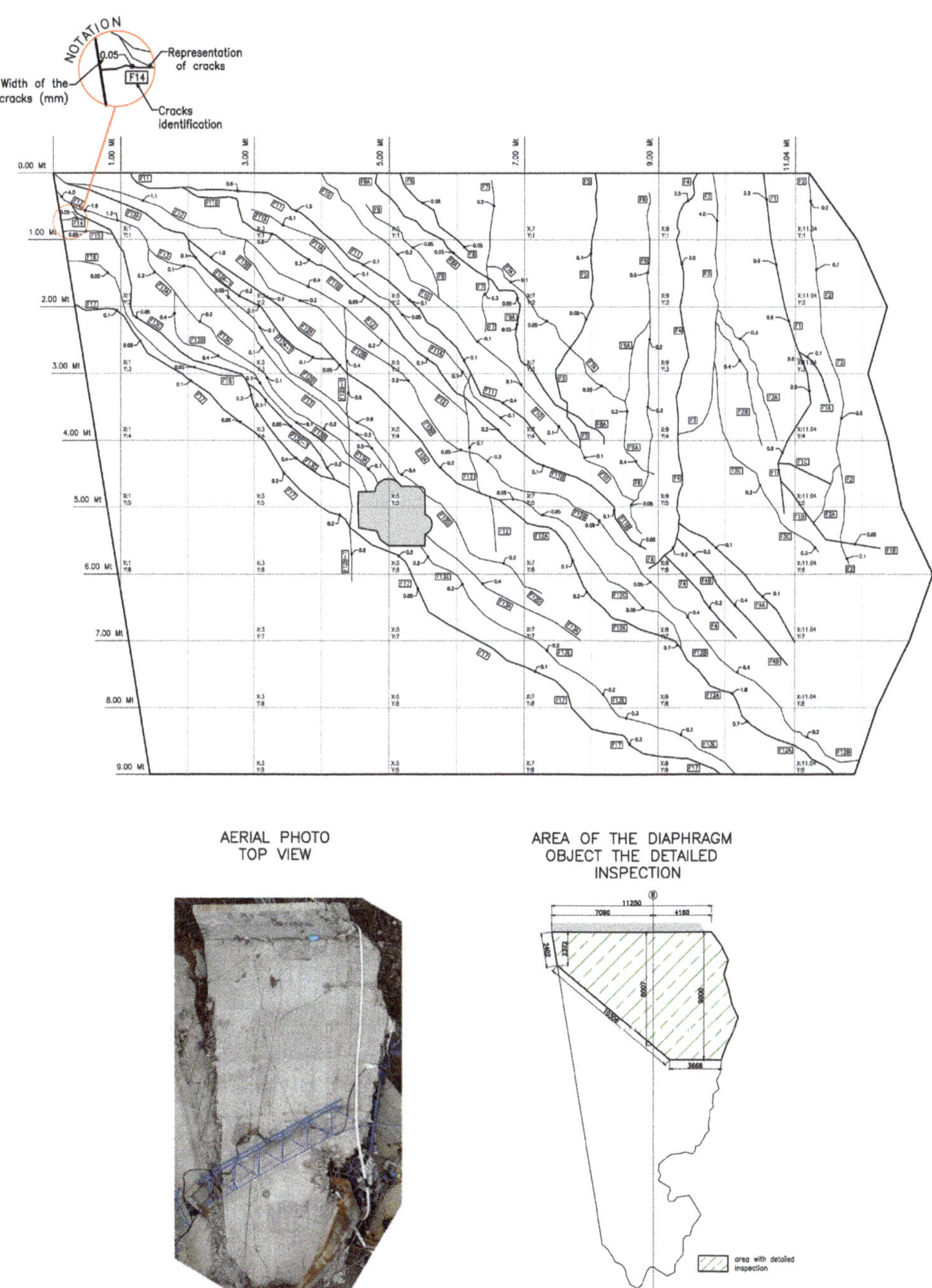

Fig. 5.14 Crack mapping of the Pylon B diaphragm after collapse

5.1.3 Pylon Head

The pylon head was observed to be largely intact, except for major damage by concrete crushing and bent and ruptured reinforcement where it was connected to the upper pylon legs (see *Fig 5.15* and *5.16*). Minor damage and concrete spalling were found on the W/E faces of the pylon head where the stay cables passed through and exited the concrete walls.

High-resolution aerial photos of the south face revealed a distinct vertical crack about 1 m from the east face of the pylon head, extending from the anchorage of the lowest stay cable to the highest stay cable (see *Fig 5.18* to *5.19*).

It was not possible to inspect the inside of the pylon head to evaluate the conditions and possible damage to the stay cable anchorages or the inside steel box.

Fig. 5.15   Pylon head base previously connected with the upper pylon legs

Fig. 5.16   Pylon head largely intact, situated east of the caisson cap on top of the girder

Fig. 5.17 Aerial photo of pylon head south face showing all installed stay cables still anchored within the pylon head

Fig. 5.18 Close-up of pylon head south face showing clear vertical crack 1 m from east face

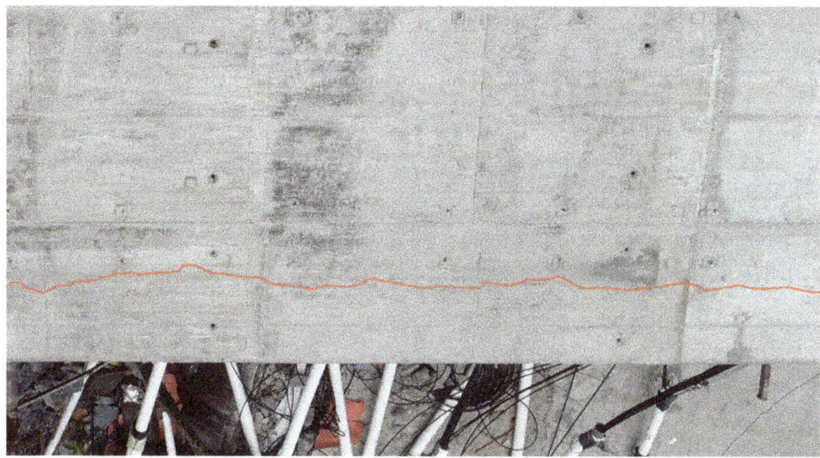

Fig. 5.19 Close-up of pylon head south face with indication of crack

5.1.4 Link Slab

Severe damage was observed on the southern part of the link slab (see *Fig 5.20* to *5.22*). This damage can most likely be explained as a result of the pylon head impacting onto the southeast part of the link slab during collapse. The significant number of longitudinal tendons resulted in the southwest part of the link slab also breaking apart. Only minor damage and superficial concrete spalling were found for the remaining part of the link slab.

Fig. 5.20 Bottom west face of the link slab with damage only apparent near the southern end of the slab

Fig. 5.21 Closer view of the damage to the southern part of the link slab

Fig. 5.22 Photo showing the top face of the link slab and the damage to the southern part of the slab

As specified in the design drawings, no connecting re-inforcement was observed between the link slab and the pylon knees (see *Fig 5.23*). Only the transverse tendons protruded from the edge face of the link slab. The concrete surface was found smooth where the link slab had separated from the pylon knees, indicating that the structural components were cast individually, resulting in a *cold* construction joint.

Fig. 5.23 Link slab north face. Smooth concrete and absence of protruding reinforcement indicating a cold joint

Fig. 5.24 Photo of the transverse tendon protruding from the link slab, before any strands
were cut and removed for testing

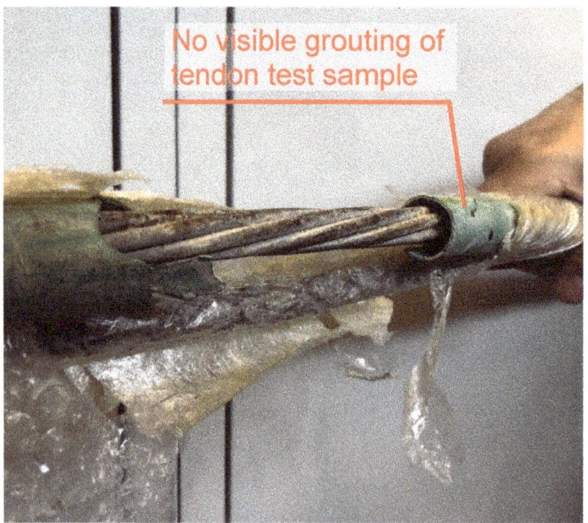

Fig. 5.25 Transverse tendon sample for material
testing. White surface layer indicating
possible grouting, though no grouting was
found when the duct was removed

Fig. 5.26 Rupturing of the transverse tendons near the
southern anchorages

The tendons were found to have ruptured close to their anchorages in the southern
pylon knee (see *Fig 5.26*), while the remaining part was pulled out of the northern edge
of the link slab, still anchored within the northern pylon knee (see *Fig 5.24*). The site

inspection showed no signs of grouting of the tendons, while the cut samples taken for testing could indicate grouting based on the surface color of the tendons, though no grouting was found when the surrounding duct was removed (see *Fig 5.25*).

5.1.5 Transition Beam

At some point during the collapse, the transition beam separated from all four anchor beams, impacted onto the ground, rotated, and slid several meters away from the abutment, with the west face turned upwards (see *Fig 5.27*). The transition beam was also found to be separated from the girder. From the separations, three types of failure were observed:

- Failure of the welding between southern anchor beams and the steel end plate (see *Fig 5.29* and *5.31*).
- Pull-out of shear studs from transition beam at the connection to the anchor beams (see *Fig 5.28* and *5.30*).
- Failure of the welds connecting the shear studs to the steel end plates. This was observed both at the connection to the anchor beam and at the connection to the girder (see *Fig 5.34* and *5.35*).

Site inspection showed indication that medium carbon threaded steel bolts of grade 5 (class 8.8 equivalent) were used in place of shear studs, welded to the steel end plates (see *Fig 5.32* and *5.33*). This type of bolt is generally considered unfit for welding.

Fig. 5.27 Transition beam detached from the anchor beams with the west face turned upwards

Fig. 5.28 Inclined anchor beam exiting Anchor Block D. Steel end plate bent with several shear studs being pulled out of the plate

Fig. 5.29 Southern anchor beams exiting Anchor Block D with indication that the end plate separated in the welded connection

Fig. 5.30 Southern connection between transition beam and anchor beams with detachment partly due to shear stud pull-out failure and partly through failure of the shear stud welds

Fig. 5.31 Southern connection between transition beam and anchor beams with end plate embedded partly into the transition beam and failure of the weld between end plate and anchor beams

Fig. 5.32 Shear studs observed on the southern end plate connecting to the anchor beam. Markings on shear studs indicate use of Grade 5 (class 8.8 equivalent) medium carbon steel bolts generally considered unfit for welding

Fig. 5.33 Shear stud welded to end plate. Threaded end indicates that bolts were used in place of shear studs

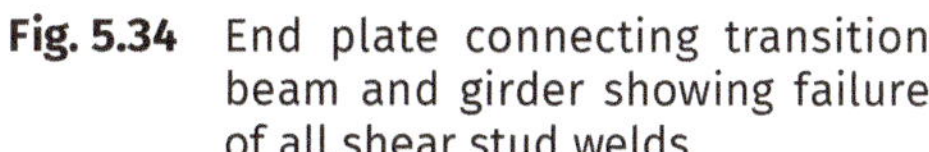

Fig. 5.34 End plate connecting transition beam and girder showing failure of all shear stud welds

Fig. 5.35 Shear studs still embedded in transition beam with failure of all welds on the connecting end plate

5.2 Pylon C and Abutment D

The following photographic records concern Pylon C, Abutment D, and the respective superstructure, which were identical to the collapsed Pylon B. All photos were taken after the collapse on January 15, 2018.

5.2.1 Cable Stays and Girder

Near the lower anchorages of anchor stays, T8DN/S to T11DN/S, all stays were observed to be situated off-center from their respective duct, away from the bridge span (see *Fig 5.36* and *5.37*). This can either be contributed to faulty placement of the ducts during construction, or to a backwards movement of the anchor beam within the anchor block consistent with the deficiencies identified in Section 2.1. A similar off-center position could be observed for several of the main-span stays (see *Fig 5.38*).

From site inspection, it was evident that neither the proposed cornices nor the recommended baffles were installed on the girder during the erection of the bridge (see *Fig 5.38*).

Below the side span, between axis C and D, one of the temporary supports for the girder was found not being removed. The northern longitudinal edge beam was seen resting on this support (see *Fig 5.39*).

Fig. 5.36   Southern anchor stays exiting Abutment D off-center from duct (T8DS, T9DS, T10DS)

Fig. 5.37   Anchor stay T11DN off-center from duct

Fig. 5.38 North upper leg just above knee and stays T1CN to T7CN. Stays off-center and no cornices on outside of longitudinal edge beams. (photo provided by GISAICO)

Fig. 5.39 North longitudinal edge beam of the girder resting on a temporary support

Fig. 5.40   Transition beam, north face, Abutment D (outlining; see Fig 5.41a)

(a)   (b)

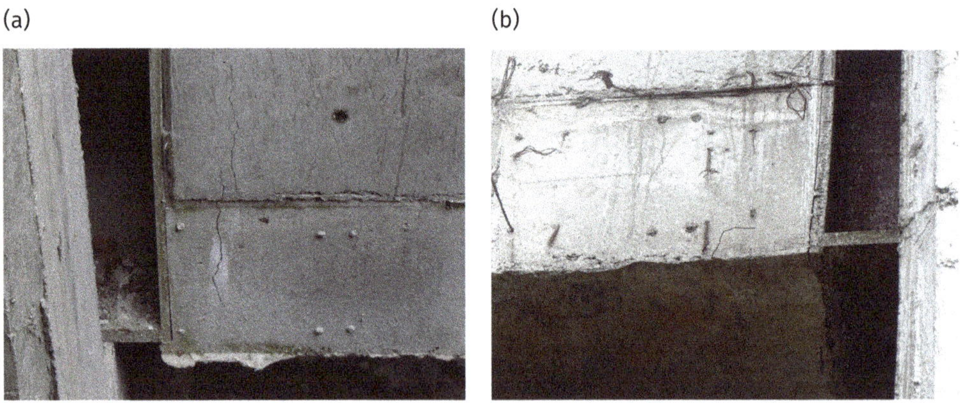

Fig. 5.41   Close-up: Cracks in lower part of transition beam at Abutment D

5.2.2 Transition Beam

It was found, that the concrete had begun to separate from the connecting steel plates on both the north and the south face at the lower end connection from transition beam to anchor beam (see *Fig 5.40* and *5.41*). A significant vertical crack had formed on the north face at a distance from the edge corresponding to the length of the applied shear studs. The crack width has been estimated to ~2 mm at its widest, by using a known reference length in the same plane as the crack. The opening between concrete and the connecting steel plate was estimated to ~3 mm. This crack formation can be related to the insufficient

Fig. 5.42 North anchor beam resting on the wall of the abutment opening

development length between the short shear studs and reinforcement, allowing for the unsatisfactory transfer of tensile stresses between the two components, identified as a deficiency in Section 2.2.

According to the construction drawings, the anchor beam should have a 50 mm gap to the concrete wall below, but it was found that the end of the anchor beam was resting on top of this wall (see *Fig 5.42*). Similarly, the anchor beam should have had 50 mm clearance to the above concrete face, but here the gap was found to be ten times as large as specified (see *Fig 5.40*).

5.2.3 Abutment Bearings

A significant shift of the neoprene bearings was observed, as they were found to not be properly aligned with the steel wedges (see Fig *5.43* and *5.44*). The shift was primarily identified in the longitudinal axis (bridge axis), but a lateral displacement was also present for some bearings (see *Fig 5.45a*). It was not possible to determine whether the bearings displaced when the cable stays were tensioned or if they were not aligned initially. One of the bearings showed the initiation of damage and cracking along one edge (see *Fig 5.45b*).

Several timber battens were observed inside the anchor block, connecting the anchor beam and the top slab, providing some level of vertical support. Presumably, these battens were used to support the above steel plate, used as formwork for the top slab of the anchor block, but were never removed (see Fig *5.43* and *5.44*).

Fig. 5.43 Neoprene bearings not aligned with steel wedges inside southern chamber of Anchor Block D

Fig. 5.44 Neoprene bearings not aligned with steel wedges inside northern chamber of Anchor Block D

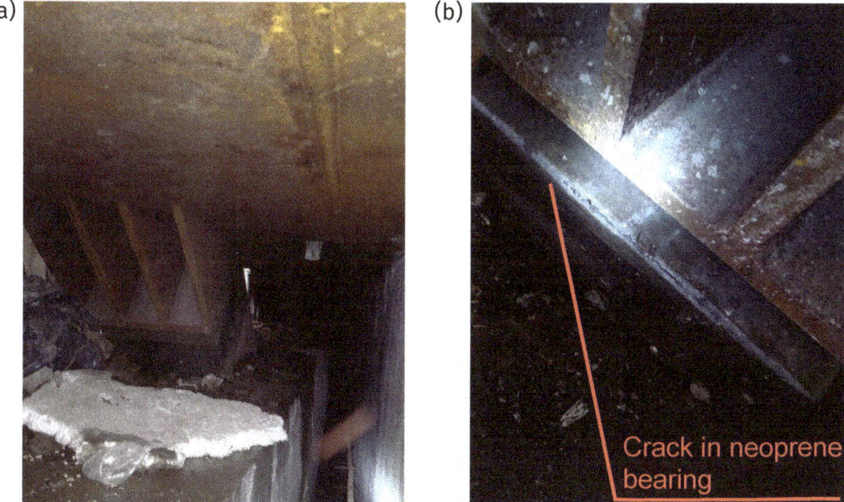

Fig. 5.45 Neoprene bearings, southern chamber, Anchor Block D
a) two-directional shift of bearing
b) cracking and damaging of bearing

5.2.4 Pylon Head

High-resolution drone photography revealed indication of a vertical crack pattern on the head of Pylon C (see *Fig 5.46*), similar to what was observed for the collapsed Pylon B head, described in Section 5.1.3. The vertical crack has formed approximately 1 m from the west face of the pylon head, consistent with the pylon head deficiency identified in Section 2.3.

Fig. 5.46 Vertical crack of Pylon C tower head
a) Identification of vertical crack
b) Vertical crack highlighted

5.2.5 Lower Pylon Legs and Diaphragm

An outwards movement of the pylon knees could be indicated by a gap between the knees and link slab, most noticeable in the upper southern joint (see *Fig 5.47* and *5.48*).

Crack formation in the diaphragm was also observed on site (see *Fig 5.49* to *5.51*). Starting from the upper northern corner on the east face of the diaphragm, a critical crack of substantial width had formed along the concave corner between diaphragm and lower pylon leg. Using the height of the link slab as known reference length, the crack width was estimated to ~5–7 mm where it was widest. The crack extended downwards from the corner approximately 4.5 m.

Fig. 5.47   East face of Pylon C diaphragm and link slab. (Outlining see Fig 5.48)

(a)

(b)

Fig. 5.48   Joint between pylon leg knee and link slab
a) Southern side.
b) Northern side. (photos provided by GISAICO)

Fig. 5.49 East face of Pylon C diaphragm including critical crack. (Outlining see Fig 5.50 to 5.52)

Fig. 5.50 Close-up of critical crack in upper northern corner of Pylon C diaphragm

Fig. 5.51 Close-up and indication of critical crack in upper northern corner of Pylon C diaphragm

Crack formations were also present in the upper southern corner of the diaphragm (see *Fig 5.52*). The crack pattern here could be described as more discrete and not with one continuous crack as in the upper northern corner. The width of cracks was also smaller, estimated at the widest point to ~1–3 mm.

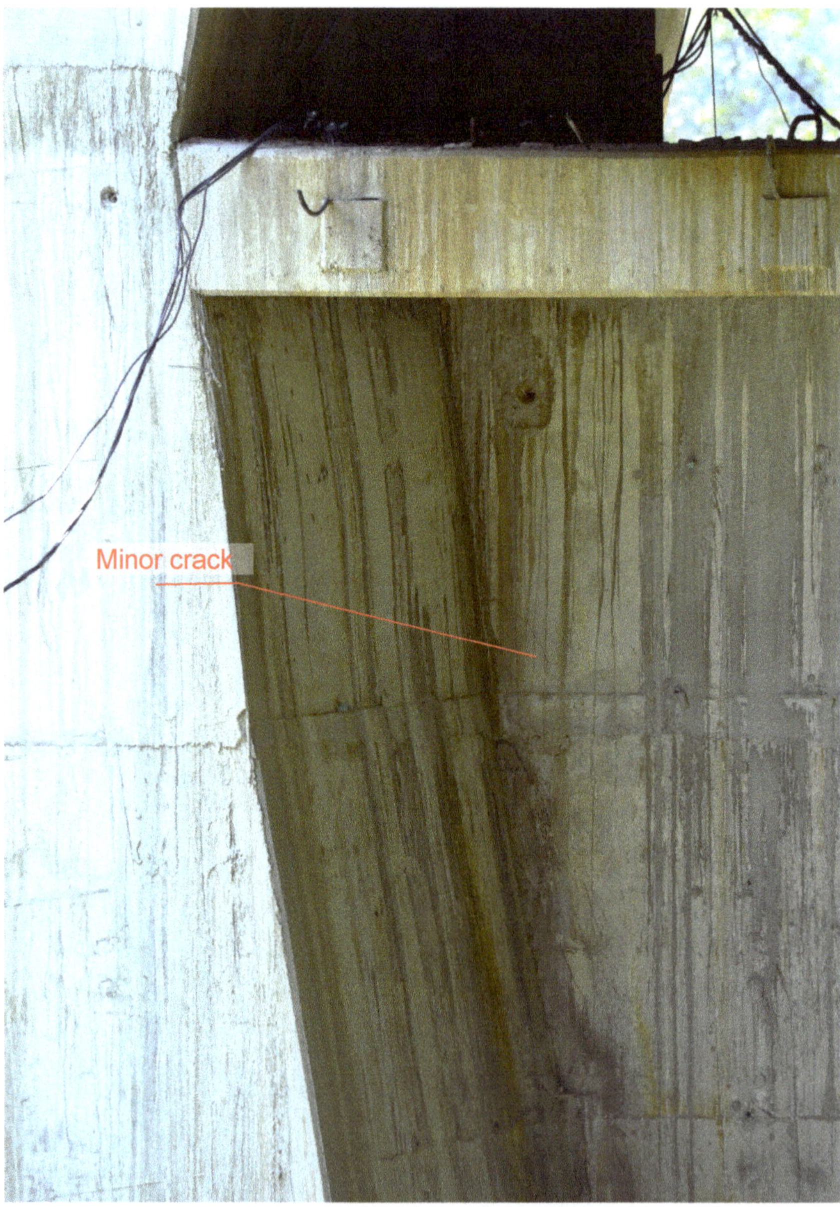

Fig. 5.52   Close-up of crack formation in upper southern corner of Pylon C diaphragm. (Photo provided by GISAICO)

Chapter 6

Interviews

Participants involved in the design and construction of the Chirajara Bridge were interviewed to clarify several issues. The general design philosophy behind the bridge as well as the construction sequence and methods were of particular interest. The interviews took place in Colombia between March 21 and 23, 2018. Excerpts from these are presented in this section.

6.1 Interview with the Main Contractor

March 21, 2018, Fabio Forero from the main contractor, CONINVIAL, was interviewed to clarify the construction sequence and to provide information regarding various issues, which turned up during construction. Excerpts from Mr. Forero's statements follow.

A drill was initially used for removal of soil from the excavation for the caissons, but when rocks became harder, dynamite was used.

The initial design depth of Caisson B was 30 m. During excavation, the former constructor, TRADECO, wanted to go deeper to find better soil conditions. The caisson was eventually founded at a depth of 34.4 m. However, here the soil was found to be in the same condition. Mr. Forero stated that the former designer, E.D.L. Ltda., specified that the foundation should not be modified any further, as it had already been evaluated by many specialists.

The link slab was not poured monolithically with the pylon legs. Only the twelve tendons in the link slab went through the pylon legs. Mr. Forero showed a construction photo of this, depicted in *Fig 6.1*.

During casting of the upper pylon legs, temporary horizontal steel tubes were installed between these to avoid big internal moments in the pylon legs until they were connected at the top. Mr. Forero showed a construction photo of this, depicted in *Fig 6.2*.

A balanced cantilever construction process of the girder was originally the idea of the former constructor, TRADECO. However, when GISAICO took over the construction work, this method was changed to a temporary supported girder, going from the anchor block to the pylon. This change was made by GISAICO to save time. The design engineer agreed with this change and redid all the calculations to approve this. *This statement contradicts a statement later made by the designer (see Section 6.2), and no engineering design calculations regarding the construction phase have been provided to the team.*

Fig. 6.1 Construction photo of non-monolithic link slab

Fig. 6.2 Construction photo of temporary steel tube, supporting upper pylon legs during construction

According to Mr. Forero, the cable supplier VSL was aware of the construction method of the girder. Cables were installed after the erection of each of the main-span sections. The temporary supports of the side span were released simultaneously with cable installation. *This statement was later confirmed by VSL.*

One temporary support at the side span of the Villavicencio side was not removed during the cable installation. Thus, a part of the girder was still supported on this at the time of collapse of the Bogota side structure. This support was not removed because it was made from concrete and not steel like the other temporary supports.

Fig. 6.3 Temporary girder support on Villavicencio side, currently still active

It was confirmed by Mr. Forero, that the temporary support at the Villavicencio side structure on *Fig 6.3* was still active at the time of the collapse of the Bogota side structure. Mr. Forero believed that this could possibly be one of the reasons why the Villavicencio side structure did not collapse.

Mr. Forero was witnessing the collapse from the Villavicencio side. It happened totally instantaneous without any warnings. No movements were felt just prior to the collapse.

6.2 Interview with the Design Engineer

March 21, 2018, Héctor Urrego, the responsible design engineer from AREA INGE-NIEROS, was interviewed to explain the design philosophy behind the bridge and to clarify questions related to the design of the bridge. Excerpts from Mr. Urrego's statements follow

The general design of the bridge was formed by considering three main issues:

1. The topography
2. The geology
3. The seismicity

The very steep topography was the reason for designing the bridge with a large main span and two smaller side spans.

A cable stayed bridge was chosen to ensure a relatively light bridge, bearing in mind the complex geology and the seismicity of the area.

The supporting neoprene pads at the abutment and on top of the link slab were chosen to ensure a flexible structure in case of earthquake. The neoprene pads at the abutment were designed to allow for both longitudinal and lateral movements.

The diaphragm was implemented to achieve a unified I-section of the lower pylon. This would heighten the lateral stiffness of the pylon for lateral wind loads and seismic

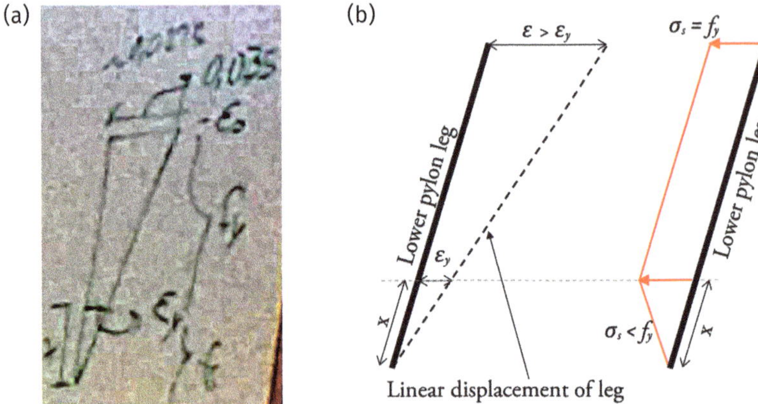

Fig. 6.4 Design method of diaphragm reinforcement, relying on ductile performance of the diaphragm
a) Sketch made by design engineer
b) Reproduction of sketch

loads. However, the main idea of the diaphragm was to absorb the horizontal deviation forces. The design engineer explained the design method of the horizontal reinforcement of the diaphragm by a sketch, (see *Fig 6.4a*). *For better understanding, this sketch was reproduced (see Fig 6.4b). The design relied on ductile performance of the diaphragm, assuming a linear deformation curve of the lower pylon legs. However, the assumption of a linear outwarded displacement of the lower pylon leg has been identified as a faulty assumption as the diaphragm would cause bending in the lower pylon leg resulting in a non-linear deformation shape activating only the upper part of the diaphragm (see Fig 2.8, Section 2.3). In addition, it was found that a relatively brittle behavior of the diaphragm could be expected due to an under-reinforced section (see Section 9.3.3).*

The main idea of the link slab was not to absorb horizontal deviation forces, but it was designed for construction reasons. The initial idea was to erect the girder by a balanced cantilever method, so that the link slab was implemented to act as a support for the girder in the construction process. This was also the reason for the 14 cables of 8 strand $\varnothing\,0.6"$ tendons in longitudinal direction and only 12 cables of 1 strand $\varnothing\,0.6"$ in transverse direction of the bridge axis.

The transition beam was implemented due to a geometrical problem in transferring compression forces from the girder to the anchor block. According to the design engineer, the transition beam was designed to resist the rotating moment, caused by this eccentric loading. *It may be noted that deficiencies related to the transition beam were found by the team in the assessment of the bridge design (see Section 2.2).*

No thorough analysis of the bridge during construction was made. It was considered unnecessary because the final stage was considered more critical than the construction stages. However, inclined safety cables were installed to give lateral stability in the cantilever construction phase. These temporary stability cables were found on drawing 039-12-S4A-DET-PTE-NC-R0, depicting two cables on either side of the bridge, consisting of 19 and 9 strands respectively, all strands tensioned to 2t. *Construction photos indicate that not all strands were tensioned (see Fig 2.21, Section 2.9), and only 14 single strands were identified at the site visits (see Fig 5.4, Section 5.1.1).*

The design engineering company have made a check of the entire bridge after the collapse, and they did not identify any problems regarding the design of the bridge. *It may be noted that several deficiencies were identified by the team in both the design of the individual structural components and the connection between these (see Section 2).*

The impression of the design engineer was that the reinforcement in the diaphragm failed very brittlely, so that the collapse might have had something to do with insufficient ductility of the diaphragm reinforcement. According to the design engineer, another possible reason could have been a sudden movement of the soil.

6.3 Interview with the Site Engineer

On March 21, 2018, Diego Ruiz from the constructor, GISAICO, was interviewed to clarify the construction sequence and to provide information regarding various issues, which turned up during construction. Excerpts from Mr. Ruiz's statements follow.

The formwork of the pylon and diaphragm was monolithic before reaching the link slab.

The $14 \times 8 \times 0.6"$ longitudinal link slab tendons were stressed alternately. However, the transverse $12 \times 0.6"$ tendons were all stressed from the south side. This one-sided tensioning was due to practical issues as a walkway was only present at the south side of the pylon.

No unusual activity was going on at the bridge at collapse, only cleaning activities.

The temporary supports were present at collapse, but they were totally free from the bridge. They had been free for more than a month before the collapse.

6.4 Interview with the Cable Supplier

On March 23, 2018, the staff of the cable supplier were interviewed to explain the stay cable design and installation process. The cable supplier was a Colombian company, Sistemas Especiales de Construcción S.A.S, licensed by VSL. Excerpts follow.

The cable supplier only provided, installed, and tensioned the cables. They were not responsible for design calculations or the determination of stay cable tensioning forces.

The cable supplier received many different versions of the stay cable tensioning levels.

The elevation of the girder was higher than expected when the cables were tensioned to the specified forces. Several bolted connections on the girder were geometrically changed to accommodate the revised elevation of the bridge girder.

Installation of cables was done simultaneously with pouring of concrete on the respective girder section. However, tensioning of the strands was always executed three days after pouring.

The cable supplier described the Chirajara project as one of the worst experiences they have ever had due to lack of information and the large number of revisions they received. Real time decisions were made through the communication system WhatsApp due to the lack of information from the beginning of the project.

Chapter 7

Geotechnical Investigation and Foundations

The team of experts commissioned Smoltczyk & Partner, Germany, to undertake an assessment of the geotechnical investigation and foundation design. This assessment has been based on information provided in design drawings, geotechnical design reports, and calculations conducted by TERRATEST S.A.S, Colombia.

7.1 Assessment of Geotechnical Investigation and Design Reports

The geotechnical investigation and design reports follow generally accepted rules of technology, in that they cover the usual scope of work for site investigations and geotechnical designs. Particularly for the abutment and Caisson B on the Bogotá side, the investigation covers:

- Consideration of topography and geomorphology
- Regional geological setting and tectonics
- Structural geology based on field mapping and boreholes
- Description of the colluvium and underlying weathered phyllite rocks in terms of their structure, discontinuities, strength, and deformational characteristics
- Hydrogeological analysis and drainage
- Subsurface geological / stratigraphic model based on boreholes and geophysical surveys
- Definition of characteristic geotechnical parameters based on rock mass classifications, laboratory tests and back calculation of existing colluvium slopes
- Numerical analysis of seepage in the slope and drainage via the caisson
- Investigation of alternative foundation types under consideration of the loads known to the geotechnical engineer at that time (e.g., $Z \approx 122\,\mathrm{MN}$ vertical load)
- Geotechnical design calculations for the chosen Caisson B (Pile 1 in the initial reports), with depth 34.5 m and diameter 8 m, including:
  - Stability of permanent slopes and construction pits
  - Stability and serviceability of foundation elements, regarding the earth pressure on the caisson shaft

    ○ Calculation of anchor forces required to stabilize caisson
    ○ Detailed recommendations for the excavation, temporary support, construction and fixing of tie-back anchors, construction of caisson structure itself

In the geological-geotechnical report on Pylon B, the geological mapping and rock mass classification of rocks exposed during the excavation of the caisson are documented to confirm the design assumptions.

On this basis, the reports document the installed instrumentation for the monitoring program during construction, fixing threshold and alarm values for the relevant deformations and reporting first observations made in the range of a few millimeters.

In a found memo, answers are given to the remarks issued by IGL (INVESTIGA-CIONES GEOTECNICAS) concerning P-Y analyses as well as numerical evaluation (by EDL) using PLAXIS 3D to evaluate the lateral deformation of the caissons with a thickness of the colluviums of 20 m at Caisson B, being on the safe side, also considering the prestressed tiebacks. Due to the observations made in the geological-geotechnical report of a colluviums thickness of only 15 m, the design was modified with appropriate n_h-factors to determine the depth-dependent subgrade reaction k_h.

7.2 Geotechnical Assessment of the Stability and Serviceability of the Foundation Caisson

As the shear parameters of the colluvium were back-calculated with a chosen global factor of safety of $F = 1.1$, additional permanent lateral support of the caisson was needed to raise the factor of safety to the required level of $F = 1.5$. This support was delivered by means of ten layers with 9 tieback anchors respectively, each pre-loaded to a working load of 500 kN. The load was calculated from the active earth pressure and alternatively by means of 3D-Finite Element Analyses, in which the forces were reduced to the special effects. German practice (EA-Pfähle) allows for the active earth pressure to be applied to a pile in a slope provided the slope stability is guaranteed for the permanent load case, which is not the case here. The geological reports also document manmade and natural slope failures in the immediate vicinity of the Chirajara valley, so that earth pressures greater than the active pressure govern.

A first back-analysis of the caisson confirms that with conservative shear strength and deformational parameters, the calculated settlements are in the order of magnitude less than 1 cm, onto which the same dimension is to be added to allow for the possible loosening of the rock after excavation. The factor of safety against bearing failure is given adequately so that there is no reason to believe that the caisson was designed to an optimized minimum.

7.3 Specific Deficiencies

Several deficiencies in the geotechnical and general design need to be addressed to prevent future problems with the caissons.

- It is not known to what extent the geotechnical design was implemented in the final design. The plans of the caisson did not show the tieback anchors as illustrated in the geotechnical design report.
- This means that many of the recommendations of the geotechnical design were apparently not heeded.
- It is not known for sure whether the micro-piles were installed under the foot of the caisson.
- Recommendations for the trimming of excess strands and caulking of the anchors have evidently not been followed. The durability of the tieback anchors can thus not be ensured when the anchor heads are exposed to the evident seepage ingress and missing corrosion protection of the strands, anchor plate, and wedges.
- No records of the specifications, drilling, grouting, and prestressing of the tiebacks have been found in the material provided. Thus, it cannot be validated whether the tiebacks were constructed as planned with the correct tendon bond and free stressing lengths in the appropriate geological layer with no abnormalities in the primary and secondary grout takes and the load-elongation lines and allowable creep rates.
- The monitoring program was not properly followed up and documented. Thus, no qualified data are available on the actual horizontal displacements and settlements at and in the vicinity of the caisson.

7.4 Summary of Geotechnical and Foundation Assessment

The site visit indicated no visible signs of distress in the Caisson B. Furthermore, the assessment of the design indicated a robust foundation at the time of collapse, even if deficiencies related to the durability and quality assurance, mentioned in Section 7.3, were to be considered.

A shock loading due to a sudden significant lateral or vertical movement of the foundation would only be a plausible cause if the caissons exhibited sufficient distress, which they did not. The design should in any event be robust enough to accommodate settlements and lateral deformations during construction and the later life of the bridge. Sufficient evidence has been found in the geotechnical design calculations that deformations in the order of magnitude of several centimeters were accounted for.

The study of the design reports, plans of the caissons, the written and photographic documentation of the collapse, as well as subsequent site investigations, laboratory tests, and back calculations, conducted by TERRATEST S.A.S., confirm the assessment that there is no geotechnical or structural evidence that the failure of the pylon was caused by or triggered by unforeseen settlements, lateral movement, or induced loads through failure of tieback anchors, earthquakes, or other geotechnical factors, such as cavities in the underground.

Chapter 8

Material Sampling and Tests

Field samples were taken from the collapsed Chirajara Bridge for the purpose of determining their condition and material properties. The field samples included concrete cores, steel rebar, steel tendons, and steel plates for test at the Technical University of Denmark (DTU) in Copenhagen, Denmark. In addition, samples of concrete cores and steel rebar were sent for test at Los Andes University in Bogotá, Colombia.

8.1 Sample Extraction

Field samples were taken from the collapsed portion of the bridge during the period of March 22 to April 11, 2018. The subcontractor in charge of the field sample extraction was PROCIESTRUCTURAS S.A.S., that provided the personnel and equipment required for the work. The specific details of the structural members and locations from where the samples were to be taken were coordinated with Brincker & Georgakis ApS. As a result, 25 zones were identified in the collapsed portion of the bridge where concrete cores, steel rebar, steel tendons, and steel plates were extracted from these zones. The samples were identified with six prefixes:

- Prefix 1: Letter Q for the insurer QBE
- Prefix 2: C for testing in Colombia and E for testing in Europe
- Prefix 3: Number to identify the zone from which the sample was extracted
- Prefix 4: Letter to identify the type of sample; N – Concrete core, S – steel reinforcing bar, T – steel tendon, L – steel plate
- Prefix 5: Number of samples from zone
- Prefix 6: Letter for condition of sample

The field samples were extracted from representative parts of the bridge including the column leg (pylon leg), diaphragm, slab on diaphragm (link slab), dado de transición (caisson cap), and the main girder.

After extraction, the samples were packaged and sent to the testing laboratories in Colombia and Denmark. Due to the collapsed condition of the bridge, many of the samples, extracted from the rubble, were likely damaged or otherwise impaired for testing under prescribed conditions. All testing was aimed at achieving the code-defined conditions for materials testing. However, it must be considered that the condition of the field samples was not fully according to code specifications, due to damage incurred during the collapse of the bridge.

In Colombia, 34 concrete cores, 12 steel reinforcing bars, zero steel tendons, and two steel plates were received for testing. In Denmark, 13 concrete cores, 17 steel reinforcing bars, eight steel tendons, and two steel plates were received for testing.

The handling and packaging of the field samples, sent to the laboratories in Colombia and Denmark, were identical and it can be assumed that there are no general differences between the two sets of samples.

Testing in Colombia mainly focused on determining the compressive strength of the concrete core samples and the yield and ultimate strength and strain of the steel samples.

Testing in Denmark focused on determining the compressive strength, the elastic modulus, and the chemical and physical composition of the concrete core samples. For the different types of steel samples, the focus was on determining the stress-strain behavior, including yield and ultimate strength and strain, elastic modulus, as well as chemical and physical composition. Additionally, the dimensional extent of yielding in the steel rebar samples as received and after testing was documented.

8.2 Test Results

8.2.1 Concrete Samples

A total of 34 concrete cores were tested at Los Andes University, to determine the concrete compressive strength. A single sample contained rebar and several samples showed signs of pre-damage in the form of cracks. The compressive strength varied between 38 MPa and 69 MPa.

The concrete cores tested at DTU have been examined to determine the compressive strength, modulus of elasticity, stress-strain response, and chemical and physical composition. 13 cored samples have been received of irregular lengths and diameter. The cores were saw-cut into 14 samples for testing. Several samples contained rebar and showed signs of pre-damage in the form of cracks and chipped concrete parts as well as wood inclusions. The compressive strength varied between 32 MPa and 55 MPa, the modulus of elasticity ranged from 20 GPa to 38 GPa.

The chemical and physical analysis of the concrete showed that the concrete had an instable mix design, causing excessive bleeding, which might be due to the addition of extra water in the mixing truck and over-vibration.

A summary of the concrete samples examined can be found in *Table 8.1*.

Tab. 8.1 Summary of concrete sample results

Structural element	Zone	Test destination	Sample ID	Compressive strength [MPa]	Modulus of elasticity [GPa]	Remarks
Pylon leg, bottom south	1	Colombia	QC-01-N-50	50.5		
		Colombia	QC-01-N-F-61	54.0		
		Colombia	QC-01-N-F-62	62.5		
		Denmark	QC-01-N-F-63	55.0	21.4	
	2	Colombia	QC-02-N-F-64	78.6		
		Colombia	QC-02-N-F-65	57.6		
	3	Colombia	QC-03-N-52	55.9		
		Colombia	QC-03-N-53	55.8		
	4	Colombia	QC-04-N-59	51.0		
		Colombia	QC-04-N-F-66	40.1		
		Colombia	QC-04-N-F-67	33.7		Damaged core
		Denmark	QE-04-N-56	41.9	19.6	
		Denmark	QE-04-N-58	44.2	23.7	
Pylon leg, upper south	5	Colombia	QC-05-N-60	57.7		
Pylon leg, bottom north	8	Colombia	QC-08-N-04	63.7		
		Denmark	QE-08-N-09_1	32.0	23.2	Rebar identified in core
		Denmark	QE-08-N-09_2	42.6	27.4	
	9	Colombia	QC-09-N-01	56.6		
	10	Colombia	QC-10-N-23	52.7		
		Colombia	QC-10-N-F-68	67.6		
	11	Colombia	QC-11-N-28	57.5		
		Colombia	QC-11-N-29	56.9		
		Denmark	QE-11-N-30	52.3	30.8	
Diaphragm	12	Colombia	QC-12-N-02	63.8		
		Colombia	QC-12-N-03	64.7		
		Denmark	QE-12-N-08	51.8	31.6	
	13	Colombia	QC-13-N13	60.8		
	15	Colombia	QC-15-N-15	50.7		
		Colombia	QC-15-N-16	64.0		
		Colombia	QC-15-N-F-69	58.1		
		Colombia	QC-15-N-F-70	61.5		1#4 Rod identified in core
		Denmark	QE-15-N-20	48.5	25.8	
		Denmark	QE-15-NP-21	54.0		Too short for extensometer
Link slab	16	Colombia	QC-16-N-19	58.9		
	17	Colombia	QC-17-N-17	58.8		
		Denmark	QE-17-N-18	49.2	27.1	
Caisson cap	18	Colombia	QC-18-N-24	48.2		
		Denmark	QE-18-N-25	53.9	31.1	
	19	Colombia	QC-19-N-22	38.3		
	20	Colombia	QC-20-N-26	48.8		
		Denmark	QE-20-N-27	36.3	38.0	
Girder deck	22	Colombia	QC-22-N-48	68.6		Damaged core
	24	Colombia	QC-24-N-F-71	67.8		
		Colombia	QC-24-N-F-72	60.9		
Pylon leg, upper north	25	Colombia	QC-25-N-73	43.9		Damaged core
		Colombia	QC-25-N-74	34.0		Damaged core
		Denmark	QE-25-N-75	41.7	23.9	
		Denmark	QE-25-N-76	34.9	31.9	

8.2.2 Steel Samples

A total of 12 rebar samples were tested at Los Andes University to determine yield strength, ultimate strength, modulus of elasticity, ultimate yield strain, and ultimate strain capacity. Considering all rebars of various sizes, the yield strength ranged from 465 MPa to 524 MPa, the modulus of elasticity ranged from 173 GPa to 226 GPa, the ultimate strength ranged from 625 MPa to 709 MPa, while both ultimate yield strain and ultimate strain capacity ranged from 11 % to 22 %.

The steel rebar, tendons, and plates tested at DTU have been examined for their yield strength, ultimate strength, modulus of elasticity, ultimate yield strain, chemical composition and documentation of the yielded and un-yielded portions as received and after testing. The steel samples were of irregular size and shape and were partly yielded prior to testing. A few samples were not suitable for testing due to their insufficient length and deformed shape. Considering all rebars of various sizes, the yield strength of the steel rebar ranged from 479 MPa to 630 MPa, the modulus of elasticity ranged from 174 GPa to 249 GPa, the ultimate strength ranged from 690 MPa to 791 MPa, and the ultimate strain capacity ranged from 0.6 % (for previously yielded samples) to 20 % for un-yielded samples. Most rebar samples showed an ultimate yield strain of 12–14 %, which is relatively low in terms of ductility but complies fully with ASTM A615 requirements [11].

The examination of the rebar diameter of the steel rebar samples as received, showed that yielding occurred over 3 to 8 rebar diameters on one side of the ruptured rebar depending on the sample. Therefore, it can be expected, based on the examined samples, that the total length of rebar involved in the yielding process during collapse of the bridge was 6–16 rebar diameters adjacent to the eventual rupture location.

The chemical analysis of the steel rebar showed a relatively high carbon content, which may be an indication of the relatively low ductility observed in tests.

Of the tendon samples, one was tested individually for each of the seven wires in the tendon. The equivalent yield strength was determined from the center wire at 1780 MPa and an ultimate strength of 1938 MPa. The modulus of elasticity of the center wire was found to be 190.7 GPa with an ultimate strain capacity of 4.64 %. Due to the deformed shape of the outer six wires, only the ultimate strength was determined and ranged from 1805 MPa to 1874 MPa. The chemical analysis of the steel tendon showed that the material meets the required properties according to the German Institut für Bautechnik.

The steel plates were tested in the in-plane direction and showed a modulus of elasticity of 203 GPa, a yield strain of 0.17 %, an ultimate strength of 513 MPa, and an ultimate strain capacity of 30 %. The chemical analysis of the steel plates showed that the material cannot be clearly classified but is likely of weldable quality.

A summary of the steel samples examined can be found in *Table 8.2*.

Tab. 8.2 Summary of steel sample results

Structural element	Zone	Test destination	Sample ID	Bar No.	Yield stress [MPa]	Ultimate stress [MPa]	Modulus of Elasticity [GPa]	Ultimate yield strain [%]	Ultimate strain capacity [%]	Yield length	Remarks
Pylon leg, bottom south	1	Colombia	QC-01-S-49	#4	512	669	211.7	17 %	17 %		
	4	Colombia	QC-04-S-51	#6	480	681	187.4	15 %	18 %		
Pylon leg, upper south	7	Colombia	QC-07-S-81	#6	465	637	206.5	17 %	19 %		
Pylon leg, bottom north	8	Colombia	QC-08-S-43	#6	486	684	183.9	16 %	19 %		
	11	Colombia	QC-11-S-83	#10	476	681	204.3	16 %	20 %		
		Colombia	QC-11-S-84	#10	489	709	177.2	16 %	20 %		
		Denmark	QE-11-S-86(F)	#4	571	716	199.9	10 %	10 %	$5\varnothing_\mathrm{S}$	
		Denmark	QE-11-S-87(F)	#4	606	717	202.4	7 %	13 %	$8\varnothing_\mathrm{S}$	
		Denmark	QC-11-S-85_1	#6	479	691	210.1	14 %	16 %		Un-yielded, sample 1/2
		Denmark	QC-11-S-85_2	#6	489	693	203.7	17 %	17 %		Un-yielded sample 2/2
		Denmark	QE-11-S-88(F)	#4	521	703	207	12 %	12 %	$7\varnothing_\mathrm{S}$	
Diaphragm	12	Colombia	QC-12-S-46	#8	468	663	226.3	16 %	19 %		
		Denmark	QE-12-S-45_1	#4	534	702	209.2	10 %	10 %		Un-yielded, sample 1/2
		Denmark	QE-12-S-45_2	#4	630	705	249.9	4 %	4 %		Un-yielded, sample 2/2
	15	Colombia	QC-15-S-12	#8	475	665	183.9	18 %	22 %		
		Colombia	QC-15-S-10	#4	524	681	178.3	11 %	12 %		
		Denmark	QE-15-S-07 (F)	#4	552	706	204.4	11 %	11 %	$6\varnothing_\mathrm{S}$	
		Denmark	QE-15-S-11_1	#4	572	734	221.4	11 %	11 %		Un-yielded, sample 1/2
		Denmark	QE-15-S-11_2	#4	567	727	206.8	12 %	12 %		Un-yielded, sample 2/2
		Denmark	QC-15-S-05(F)	#4		749	174.2	1 %	1 %		Not properly tested
		Denmark	QE-15-S-14 (F)	#4	550	709	195.1	12 %	20 %	$8\varnothing_\mathrm{S}$	Necking inside extensometer
		Denmark	QE-15-S-47(F)	#4	557	736	210.8	12 %	12 %	$3\varnothing_\mathrm{S}$	
Girder	23	Colombia	QC-23-S-82	#4	474	625	173.1	13 %	11 %		
Pylon leg, upper north	25	Colombia	QC-25-S-77	#6	476	650	190.4	22 %	22 %		
		Colombia	QC-25-S-78	#6	480	662	210.1	19 %	19 %		
		Denmark	QE-25-S-79(F)	#6		774	197.3	1 %	1 %		
		Denmark	QE-25-S-80(F)	#6		791	205	1 %	1 %		
Link slab tendons	11	Denmark	QE-11-T-31_1			1813					Curved wire
		Denmark	QE-11-T-31_2			1806					Curved wire
		Denmark	QE-11-T-31_3		1780	1939	190.7	5 %	5 %		Straight wire
		Denmark	QE-11-T-31_4			824					Curved wire
		Denmark	QE-11-T-31_5			1784					Curved wire
		Denmark	QE-11-T-31_6			1874					Curved wire
		Denmark	QE-11-T-31_7			1874					Curved wire
Girder structural steel	22	Denmark	QE-22-L-37		398	489.7	203.2	15 %	31 %		
	23	Denmark	QE-23-L-35		400.4	513.2	239.5	15 %	30 %		

Chapter 9

Detailed Investigation of the Collapse of Pylon B - West

A detailed material and geometrical non-linear finite element (FE) analysis of the collapsed structure was conducted to establish an accurate collapse mechanism and to determine the main cause of collapse. A thorough description of the FE-model, as well as the outcome of the analysis, is presented herewith.

The purpose of the FE-model is to reflect the actual structure and the acting load at the time of collapse. Thus, no safety or modification factors have been applied to either the material properties or the loads. The material properties aim to reflect the actual behavior which could be expected, based on the material tests and the bridge design.

Prior to the detailed FE-model, presented in this section, a simple independent preliminary model was established by analytically derived equations and solved by numerical integration. The simple model indicates similar behavior of the structure and identical mechanisms with respect to the collapse as found by the detailed FE-model presented in this section.

The finite element software SOLVIA was used for the detailed investigation of the Pylon B collapse.

9.1 FE-Model Description

The structure was divided into different structural components. Each component as well as the connections between them were modeled to reflect the actual behavior of the overall structure.

9.1.1 Statical System

The general static system of the FE-model is seen in *Fig 9.2*, whereas the detailed modeling of boundary conditions can be seen in *Fig 9.3*.

To establish a reliable analysis, resembling the actual structure prior to collapse, the temporary bracing stays from the caisson cap to the girder were not included in the analysis. These were excluded since it was unclear how many of the strands were provided, how many were tensioned, and to which level they were tensioned (see *Fig 2.21*, Section 2.9).

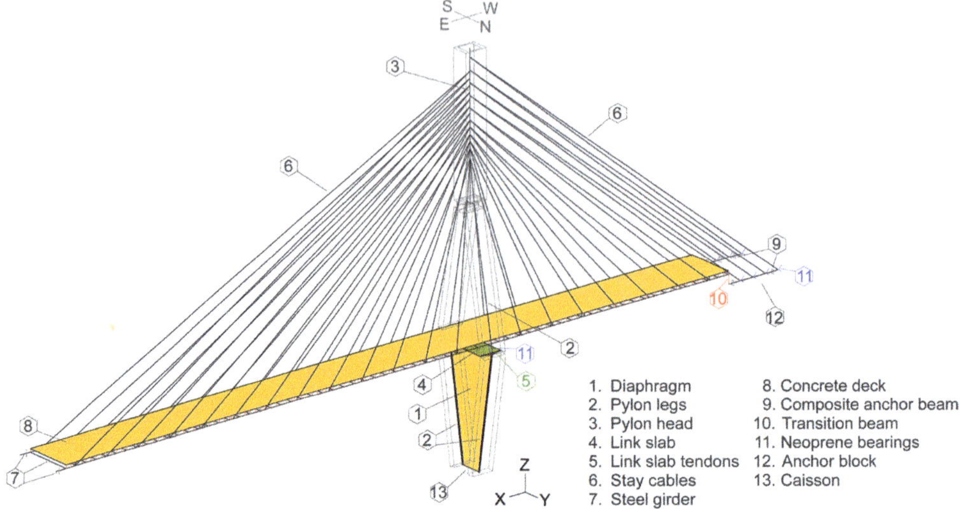

Fig. 9.1   3D FE-model of Pylon B, depicting individual components (see Tab. 9.1)

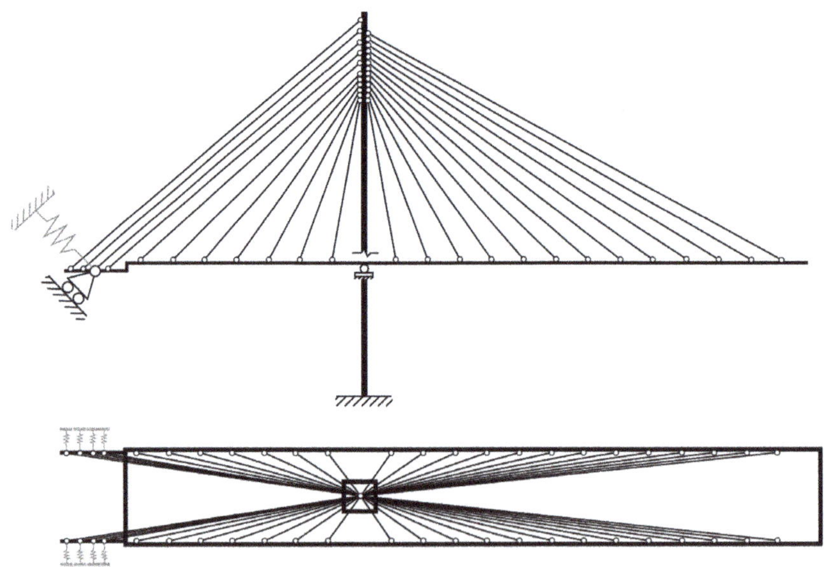

Fig. 9.2   Overview of static system and support conditions for FE-model of Pylon B

9.1.2 Structural Components

An overview of the structural components included in the FE-model can be found in *Table 9.1*. Each structural component consisted of several finite elements; the behavior defined by the element type. The node number listed in *Table 9.1* defines the number of nodes for a single element. The components are numbered from 1 to 13 and can be identified in *Fig 9.3*. Here, the structural components of the FE-model are highlighted, whereas the outline of the actual structure from the design drawing is shown shaded. Shell elements are filled and shown with a thickness, whereas beam and truss elements are illustrated by lines.

It should be noted that the modeling of the pylon legs has been subject to extensive analysis to ascertain whether a shell element model would provide better insight into the structural behavior. It was concluded that this was not the case. Thus, the pylon legs were eventually modeled by beam elements for the sake of simplicity and computational reasons.

Tab. 9.1 FE-model element description

Structural components	Element type	Element Nodes	Total Number of elements
1. Diaphragm	Shell	16	1530
2. Pylon legs	Beam	4	240
3. Pylon head	Beam	2	54
4. Link slab	Shell	16	294
5. Link slab tendons	Truss	2	26
6. Stay cables	Truss	2	48
7. Steel girder	Beam	2	402
8. Concrete deck	Shell	4	1328
9. Composite anchor beam	Beam	2	16
10. Transition beam	Rigid link	2	2
11. Neoprene bearings	Spring	2	26
12. Anchor block	Rigid support	-	-
13. Caisson	Rigid support	-	-

9.1.3 Connection of Structural Components

The diaphragm and the lower pylon legs were cast monolithically and connected by both vertical and horizontal reinforcement. Therefore, the diaphragm was modeled with rigid links connecting to the centerline of the lower pylon legs (see *Fig 9.3a*).

As can be seen in *Fig 9.3c*, the diaphragm was modeled slightly wider than the slab to ensure that tension stresses from the outward displacement of the legs were transferred through the diaphragm to the link slab. Since the link slab and the pylon legs were not cast monolithically and no reinforcement went through the connection between the link slab and the pylon knee, tensile stresses could not have been transferred to the link slab in this connection. Compression-only elements were thus implemented in the FE-model in this connection to transfer compression stresses, caused by post-tensioned transverse tendons in the link slab.

As illustrated in *Fig 9.3c*, the connection between the girder and the link slab was modeled by spring elements with stiffness in z-direction corresponding to the applied neoprene pads. Similarly, the neoprene pads between the composite anchor beam and the anchor block were modeled by spring elements (see *Fig 9.3b*). These were axial springs of high stiffness in the direction of the stays and springs of minor stiffness perpendicular to

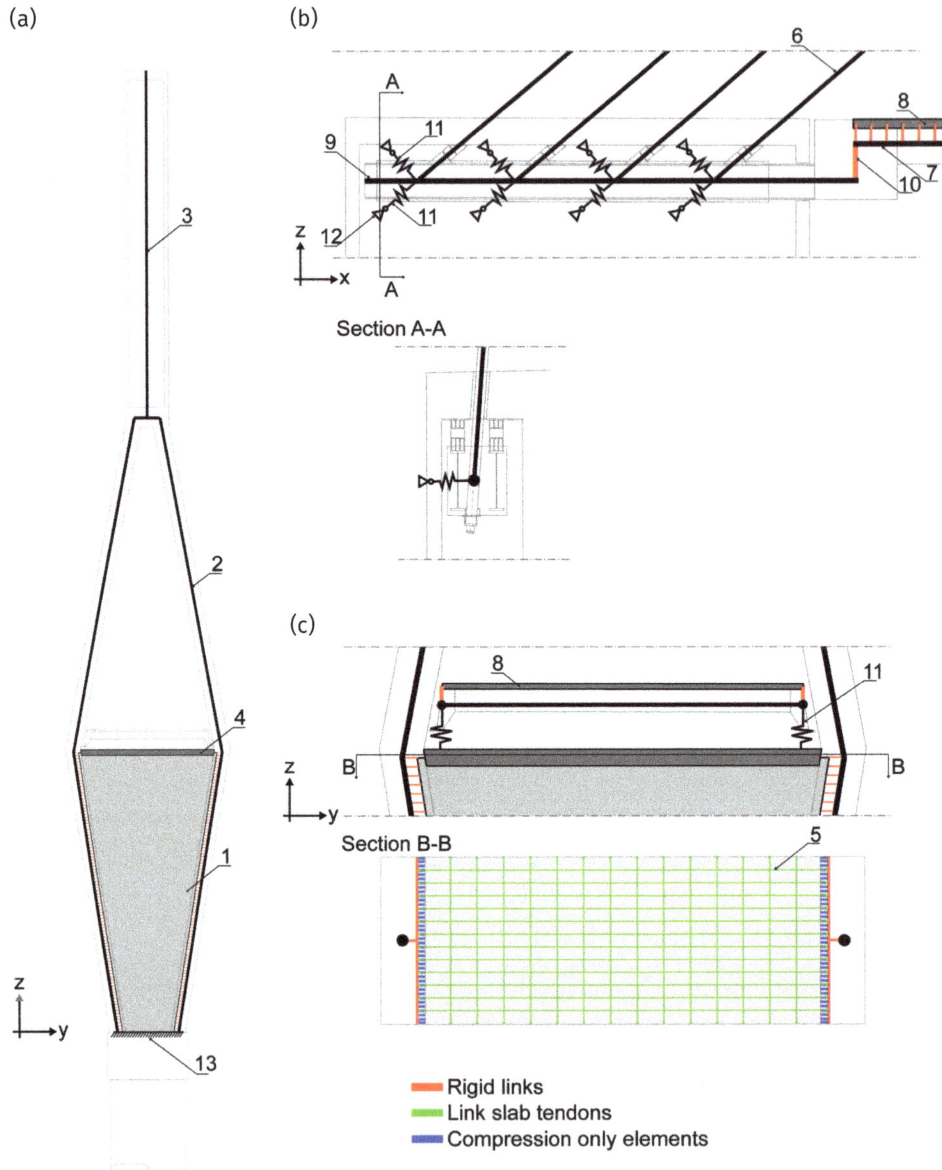

Fig. 9.3 Details of FE-model (see table 9.1)
    a) Modeling of pylon
    b) Detail of anchor support
    c) Detail at slab level

the direction of the stays, modeled to reflect the stiffness of the neoprene pads. Properties of the springs can be found in *Table 9.4*.

Shear connectors were welded to the steel girder, while rigid links were implemented in the FE-model as connections between the steel girder and the concrete deck (see *Fig 9.3b*).

9.2 Material Properties

9.2.1 FE Material Properties

Material properties reflecting the properties found from sample testing of the actual structural components were applied in the FE-model. The exact values applied are presented in *Tables 9.2* to *9.4*. Only the parameters which were used as input for the materials of the FE-model are shown.

Tab. 9.2 Concrete model material properties

Concrete	f_c [MPa]	f_{cu} [MPa]	f_{ct} [MPa]	E_0 [GPa]	ε_c [-]	ε_{cu} [-]	v [-]	γ [kN/m³]
Pylon legs	52.0	37.0	1.0	30	0.0023	0.0035	0.2	25
Diaphragm	57.0	40.0	4.4	30	0.0023	0.0035	0.2	25
Link slab	56.0	40.0	4.4	30	0.0023	0.0035	0.2	25
Pylon Head				30			0.2	25
Girder concrete deck				30			0.2	25

Tab. 9.3 Steel model material properties

Steel		$\o_s$ [mm]	f_y [MPa]	E_s [GPa]	E_h [MPa]	v [-]	γ [kN/m³]
Reinforcement	Rebar type						
Pylon legs	Vertical #10	32.3	457	200	1281	0.2	78
Diaphragm	Horizontal #4	12.7	509	200	1396	0.2	78
Link slab	Horizontal #4	12.7	509	200	1396	0.2	78
Pylon Head	Vertical #10	32.3		200		0.2	78
Cables							
Transverse tendons, link slab		15.2	1780	191	2200	0.2	78
Longitudinal tendons, link slab		15.2	1780	191	2200	0.2	78
Stay cables		15.7		200		0.2	78
Structural steel							
Steel girder				200		0.2	78
Anchor beam				200		0.2	78

Tab. 9.4 Spring element properties

Spring elements	direction [-]	k [kN/m]
Girder/slab support	z	1.22 E+06
Anchor support	Parallel with stays	4.07 E+06
	Orthogonal to stays	7.81 E+02

9.3 Material models

The material models applied to the structural components are listed in *Table 9.5*. Illustrations of the stress-strain relation of these are presented in *Fig 9.4* to *9.7*. The applied material properties presented in Section 9.2, were used as input for the material models to achieve a realistic behavior of the FE-model. Each parameter can be identified in the illustrations.

Tab. 9.5 Material model applied to structural components

Structural component	Material type	Material model
Pylon leg	Concrete	Concrete
	Rebar	Plastic
Pylon head	Concrete	Elastic
	Rebar	Plastic
Diaphragm	Reinforced concrete	Nonlinear
Link slab	Reinforced concrete	Nonlinear
Girder concrete deck	Reinforced concrete	Elastic
Tendons, link slab	Steel cable	Plastic
Stay cables	Steel cable	Tension only
Steel girder	Structural steel	Elastic
Anchor beam	Structural steel	Elastic

The pylon leg and pylon head were modeled by two material models, representing the concrete and the reinforcement respectively (see *Fig 9.4* and *9.5*). By combining these two material models, with the reinforcement included in the correct positions, the elements behaved as reinforced concrete elements.

Tendons in the link slab were modeled by the bilinear plastic material model, allowing for redistribution of forces into the diaphragm.

The steel girder, anchor beam, and concrete girder deck were modeled by the elastic material model, as deformations in these components purely occurred in the elastic range of the respective materials (see *Fig 9.6*).

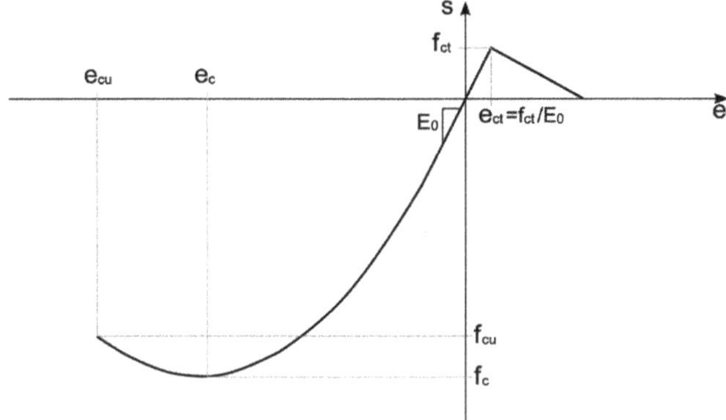

Fig. 9.4 Stress-strain relation for concrete material model

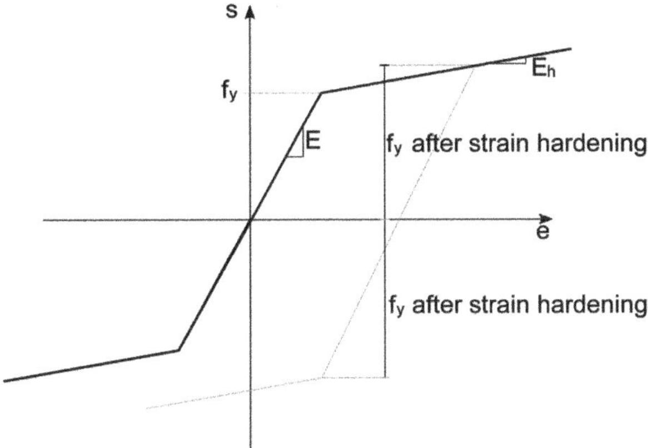

Fig. 9.5   Stress-strain relation for plastic material model

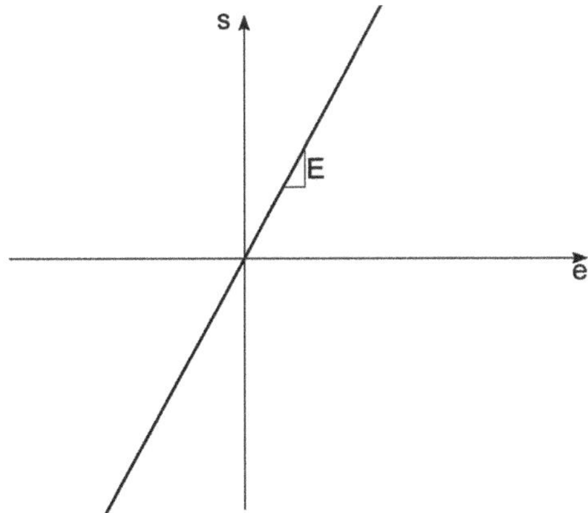

Fig. 9.6   Stress-strain relation for elastic material model

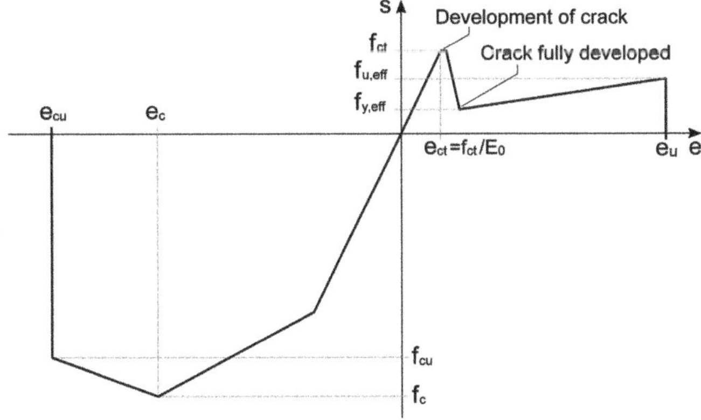

Fig. 9.7   Stress-strain relation for non-linear material model

Stay cables were modeled with a bi-linear tension-only stress-strain relationship and would exhibit no stresses for compressive strains.

The diaphragm was modeled by one single non-linear material model, which represents both the concrete and the reinforcement. Hence, the contribution from the reinforcement in the diaphragm was accounted for by an effective equivalent stress.

9.3.1 Ultimate Capacity of Pylon Legs

Ultimate strain limits were defined in the FE-model for the concrete material model (see *Fig 9.4*). However, the rebars included in the pylon legs followed a bilinear stress-strain relation allowing for infinite strains. Thus, the actual combination of forces was checked in the FE-model results to detect the point of rupturing of the pylon legs.

Failure surfaces for different values of axial forces combined with two-directional bending of the pylon legs were evaluated (see *Fig 9.23*).

Ignoring contributions from concrete aggregate interlock and dowel action, considering only the 15 horizontal N°4 stirrups, spaced at 200 mm and a value of $cot\,\theta = 2$, the shear capacity was found to be:

$$V_S = \frac{A_s}{s} f_y \, z\,cot\,\theta \approx 12MN \tag{9.1}$$

9.3.2 Ultimate Deformation Capacity of Tendons and Link Slab

Just as for the reinforcement, the material model applied to the tendons in the link slab allowed for infinite strains in the FE-model. Thus, the actual strain in these was checked in the FE-model results to detect the point of rupturing of tendons. The transverse tendons were essentially ungrouted, resulting in little to no bonding or stress transfer to the surrounding concrete. As the transverse tendons were anchored in each end, at the north and south knee respectively, the tendons were assumed to be able to reach their full strain capacity as the knees would move apart. Besides the transverse tendons, the outwards movement of the knees was restrained only by the diaphragm and the flexural stiffness of the pylon legs. The link slab was cast separately from the legs in a *cold* joint with no continuous reinforcement. Thus, this connection provided no restraint towards an outwards movement, as tensile stresses could only be transferred from the pylon legs into the link slab through the diaphragm.

As the outer faces of the knees were 17600 mm apart, subtracting 300 mm at each end for anchoring, the effective length of the tendons can be considered ~17000 mm. Depending on the chosen ultimate strain of the strands, the deformation capacity as allowable outwards movement of each knee is **~250–600 mm** given **3.0–7.0 %** ultimate strain before rupture of the tendons commences.

9.3.3 Ultimate Deformation Capacity of Diaphragm

The upper part of the diaphragm was exposed to tension, as this component partly absorbed the horizontal deviations forces, induced from the kink of the pylon legs. Ductile performance can be expected when considering an unconfined reinforcement bar. However, the average strain capacity is significantly reduced for a reinforced concrete member in tension, as strains are concentrated in cracks. This reduction is amplified when considering small reinforcement degrees. When considering an under-reinforced section, a very brittle behavior with almost zero deformation capacity can be expected, when the full width of the diaphragm is considered [2–6].

The behavior of the reinforcement in tension that connected the upper part of the diaphragm to the lower pylon legs can be assessed analytically by considering a simplified tension chord, illustrated in *Fig 9.8*, in both the pre- and post-yield response based on the findings of Fernández Ruiz et al [1], extending on the tension chord model proposed by Marti et al [3].

In the case of uniaxial tension, the bond response can be derived from the governing differential equation ensuring equilibrium:

$$\frac{\pi}{4}\phi_s^2 d\sigma_s = -\tau\pi\phi_s\, dx \tag{9.2}$$

For fully anchored bars, assuming affinity of the bar slip, the differential equation in eq. (9.2) can be solved in closed-form by adopting a specific relation between bond and slip. Proposed by Fernández Ruiz et al [1], the best agreement with experimental results is found from the 'square root model' assuming a third-root relationship between bond and slip ($\tau \propto \delta^{1/3}$), resulting in a square root relation between bond and reinforcement steel strains, hence the name. The bond parameters can be determined, based on the compressive strength of concrete, according to *fib* Model Code 2010 [13].

Considering the decrease of bond, due to contraction of the rebar as well as the phenomenon of local punching by conical cracks radiating from the reinforcement ribs close to the primary crack [7], the rebar steel strain can be expressed as a function of distance x from the primary crack.

$$\varepsilon_s = \begin{cases} \left(\sqrt{\varepsilon_y} - \dfrac{2\tau_{b,\max}\left(x-l_p\right)}{E_s\phi_s\sqrt{\varepsilon_y}}\right)^2, & x > l_p \\[2em] \varepsilon_{bu} - \left(\varepsilon_{bu} - \varepsilon_y\right)\exp\left[\dfrac{4\tau_{b,\max}}{E_h\phi_s\left(\varepsilon_{bu}-\varepsilon_y\right)}\left(\left(x-l_p\right) - \phi_s\left(\exp\left[-\dfrac{l_p}{\phi_s}\right] - \exp\left[-\dfrac{x}{\phi_s}\right]\right)\right)\right], & x \leq l_p \end{cases} \tag{9.3}$$

The length l_p at which the bar is yielding can be found by rewriting eq. (9.3) in the case of $x = 0$ and $\varepsilon_s = \varepsilon_{su}$, which yields the following equation that can be solved iteratively.

$$-\frac{l_p}{\phi_s} = \ln\left[\frac{\varepsilon_{bu} - \varepsilon_{su}}{\varepsilon_{bu} - \varepsilon_y}\right]\frac{E_h\left(\varepsilon_{bu} - \varepsilon_y\right)}{4\tau_{b,\max}} - 1 + \exp\left[-\frac{l_p}{\phi_s}\right] \tag{9.4}$$

The strain capacity of the reinforced concrete member can then be calculated as the average tensile strain of the reinforcement bar along the entire length of the bar. This is usually when cracks are distributed and evenly spaced, equivalent to the average tension between two adjacent cracks. However, for an under-reinforced tension chord, the capacity will drop as a result of the first crack formation, and a second crack is unable to develop. In continuation hereof, it should be mentioned that the diaphragm fulfills the conditions of an under-reinforced section. Assuming symmetry about primary crack axes this can be expressed as:

Distributed cracking:

$$\varepsilon_{RC,u,dist.} = \frac{1}{s_{rm}} 2 \int_{x=0}^{x=\frac{1}{2}s_{rm}} \varepsilon_s(x)\,dx \qquad \text{(a)}$$

$$\text{(9.5)}$$

Single crack formation:

$$\varepsilon_{RC,u,sing.} = \frac{1}{L} 2 \int_{x=0}^{x=\frac{1}{2}L} \varepsilon_s(x)\,dx \qquad \text{(b)}$$

Thus, the strain capacity of a long member with single crack formation decreases significantly compared to a member with distributed cracking and even more so with respect to the capacity of the *naked* steel bar ε_{su}. The principle is expressed in eq. (9.6) and shown in *Fig 9.8*.

$$\underbrace{\frac{1}{L} 2 \int_{x=0}^{x=\frac{1}{2}L} \varepsilon_s(x)\,dx}_{\varepsilon_{RC,u,sing.}} < \underbrace{\frac{1}{s_{rm}} 2 \int_{x=0}^{x=\frac{1}{2}s_{rm}} \varepsilon_s(x)\,dx}_{\varepsilon_{RC,u,dist.}} < \varepsilon_{su} \qquad \text{(9.6)}$$

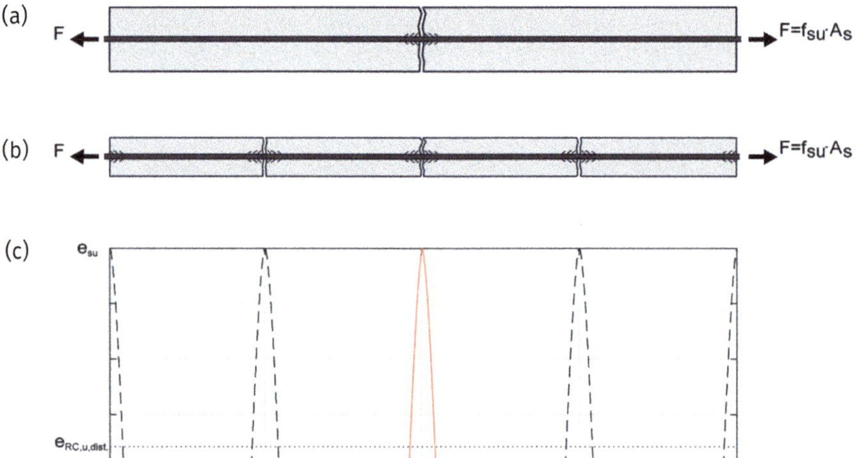

Fig. 9.8 Reinforcement strain distribution
a) Under-reinforced RC tension chord
b) Normal-reinforced RC tension chord
c) Strain distribution of under-reinforced (red) and normal-reinforced (dashed) tension chord

By increasing the finite length of the tensile member, in the case of single crack formation, the average strain capacity decreases. Considering the geometry of the diaphragm as input for the finite length, it is possible to estimate the allowable outwards movement of the knees before rupturing of the uppermost horizontal reinforcement in the diaphragm. Using the material parameters reflecting the diaphragm, the average strain capacity over the full width of the diaphragm for horizontal tensile loading was found to be ~**1‰**. Thus, the deformation capacity corresponds to a total elongation of ~**14mm** over the full width. Even though the parameters applied in assessing the deformation capacity of reinforced concrete are associated with uncertainties, this limit was used as input for the material model of the diaphragm component in the FE-model.

9.4 Loads

The loads applied in the FE-model were chosen to reflect most accurately the conditions at the time of collapse and are thus different from the load combinations used to evaluate the design in Section 2.

The wind load and construction live load were left out, as these are considered insignificant at the time of collapse. However, a small lateral load of 1/10000 of the vertical load in the pylon head was added to the top of the pylon to imitate asymmetric loading, introduced from e.g., small winds or imperfections. Besides the small lateral load, only the structural deadload of each structural component was applied in the FE-model. Densities are specified in Section 9.2, *Table 9.2 to 9.4*.

9.4.1 Load History

The FE-Analysis was separated into twenty static load stages, including no dynamic effects or inertia forces, and a following dynamic analysis. The intent of the static load stages was to capture important features due to the non-linear material characteristics. The load history started at 50 % of the structural dead load. At load stage 20, the load applied in the FE-model corresponds to the dead load of the structure at the time of collapse. Load stage 20 is also referenced as time 0.000 s in the following dynamic analysis (see *Fig 9.9*).

Relevant effects from the actual sequential construction process were accounted for in the FE-model during the static load stages. Thus, the post-tensioned link slab tendons were introduced in the FE-model under approximately the same conditions that were present at the actual time these were tensioned during construction (see *Fig 9.10*). The axial force in the upper pylon legs at knee level at the time of tensioning was estimated in eq. (9.7), simply by the weight of the inclined upper pylon legs present at that stage of construction.

$$F = (1.6 \cdot 6 \cdot 16.55)\text{m}^3 \cdot 25\text{kN/m}^3 \cdot \cos(10.8°) = 3902\text{kN} \tag{9.7}$$

The axial force at pylon knee level was 4080 kN at the time the post-tensioned link slab tendons were introduced in the FE-model. This is satisfactorily consistent with the axial force calculated in eq. (9.7) (see *Fig 9.10*).

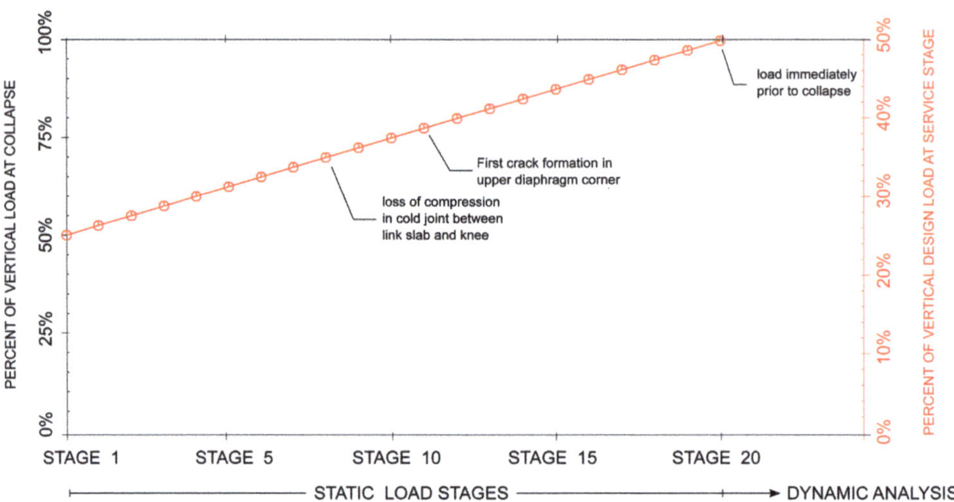

Fig. 9.9   FE-model load history. Twenty static load stages with no dynamic effects or inertia forces, followed by a dynamic analysis from stage 20, referred to as reference time 0.000

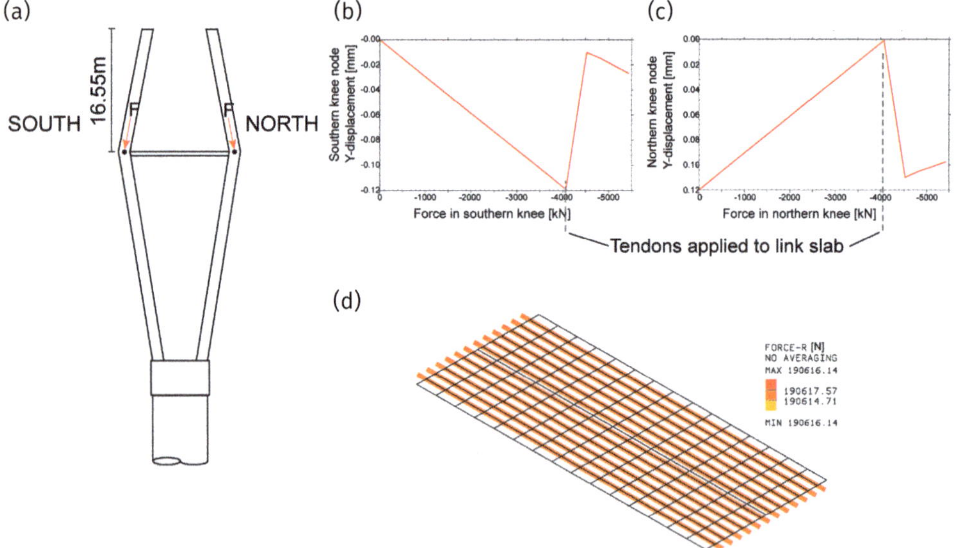

Fig. 9.10   Applying link slab tendons in FE-model
   a) Construction status of Pylon B at actual time of tensioning
   b) Southern knee displacement as function of axial force at pylon knee level
   c) Northern knee displacement as function of axial force at pylon knee level
   d) Force in post-tensioned tendons

The tendons were applied with an initial strain, such that the force in each strand reached a value of approximately 19 t (see *Fig 9.10d*). This corresponds to the specified post-tensioning force of 20 t, considering a loss of 1 t. After the post-tensioned tendons in the link slab were introduced, the load increased until the load present immediately prior to collapse was attained.

9.5 Results of Detailed Non-linear Finite Element Analysis

Initially, a description of the collapse mechanism attained from the FE-model results are given. In addition, these results are compared to the video of the collapse. Subsequently, relevant results of the analysis, including reactions, internal forces, deformations, etc. before and during collapse are presented.

9.5.1 Collapse Mechanism

Results from the FE-model analysis show a sudden separation of the diaphragm from the lower southern pylon leg to ~101 % of the load present at the time of collapse (see *Fig 9.11*).

The diaphragm rupture is a consequence of the significantly reduced stiffness contribution of the link slab occurring after contact between link slab and pylon legs vanished, whereupon the stiffness is only represented by the 12 transverse tendons. Thus, the horizontal forces at the pylon knee, induced due to the diamond-shaped geometry of the pylon, are transferred to the upper part of the diaphragm. Because of the increasing load, the lower pylon legs then needed to deform to activate a larger zone of the diaphragm to reach new equilibrium states, at the same time as the link slab tendons elongate elastically. At some point, the ultimate strain capacity in the upper south part of the diaphragm is reached, which leads to an asymmetrical sudden se-

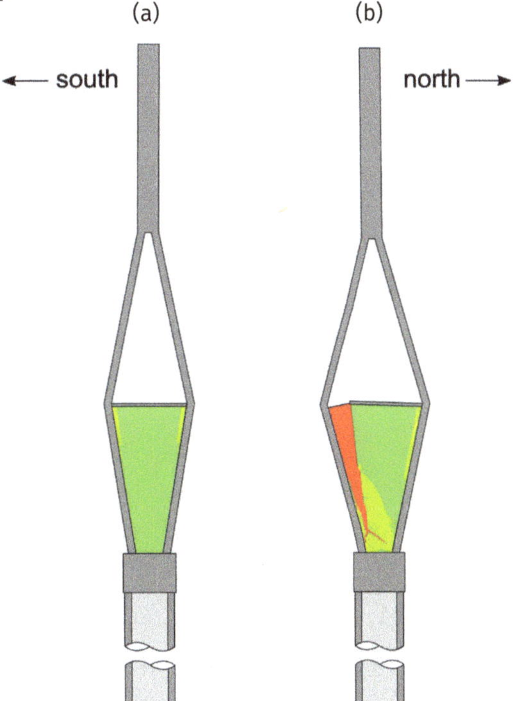

Fig. 9.11 FE-model result of sudden diaphragm rupture along the lower southern pylon leg
a) Conditions immediately prior to collapse
b) Rupture of diaphragm reinforcement followed by large displacements

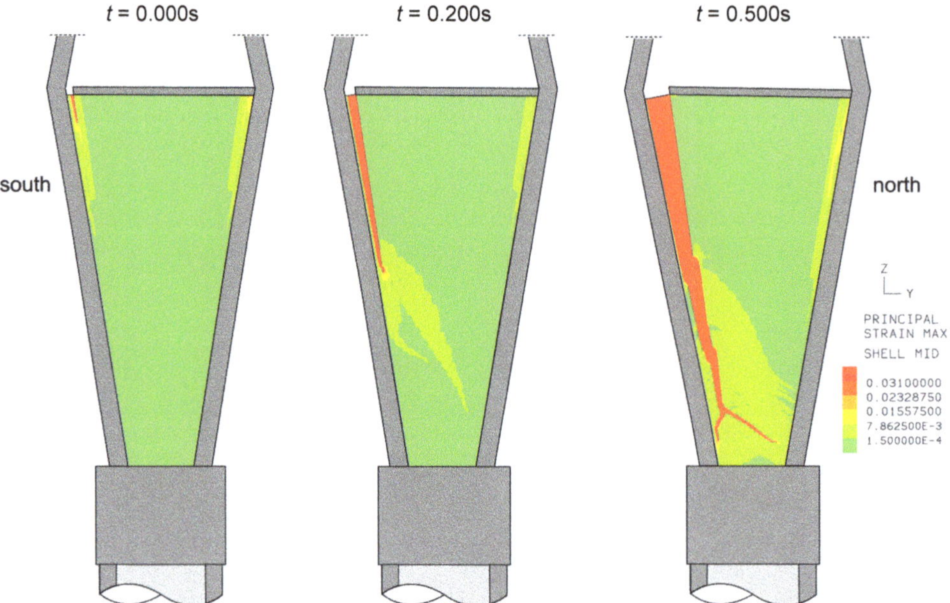

Fig. 9.12   Sequential failure formation after first diaphragm reinforcement rupture

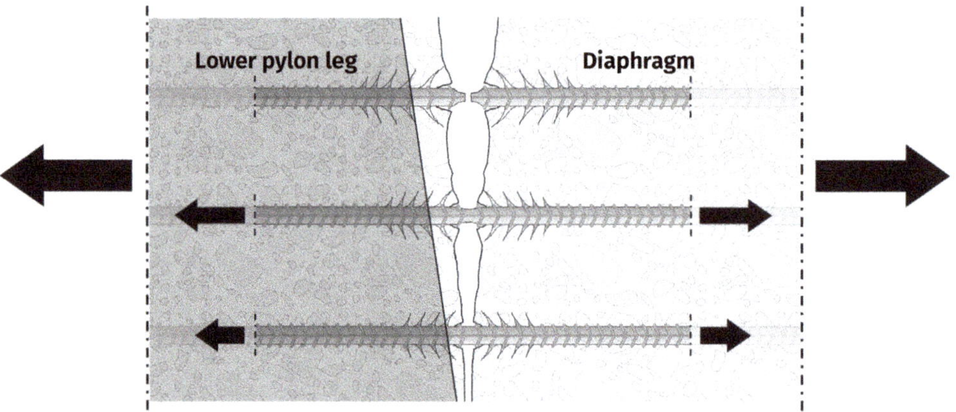

Fig. 9.13   "Unzipping" effect of sequentially rupturing reinforcement along the pylon edge of the diaphragm

quential rupturing of the reinforcement along the entire diaphragm, connected to the lower southern pylon leg (see *Fig 9.12* and *9.13*).

The brittle rupture of the diaphragm reinforcement initiates a dynamic failure mechanism. An abrupt change of the static system is followed by large outwards displacement of the knees, amplified by dynamic inertia forces. The sudden displacements of the pylon knees for this load can be seen in *Figs 9.14a* and *9.14b*.

Change of internal forces is accompanied by an outwards movement and displacement of the pylon knees, as shown in *Fig 9.14*. The displacements immediately prior to the

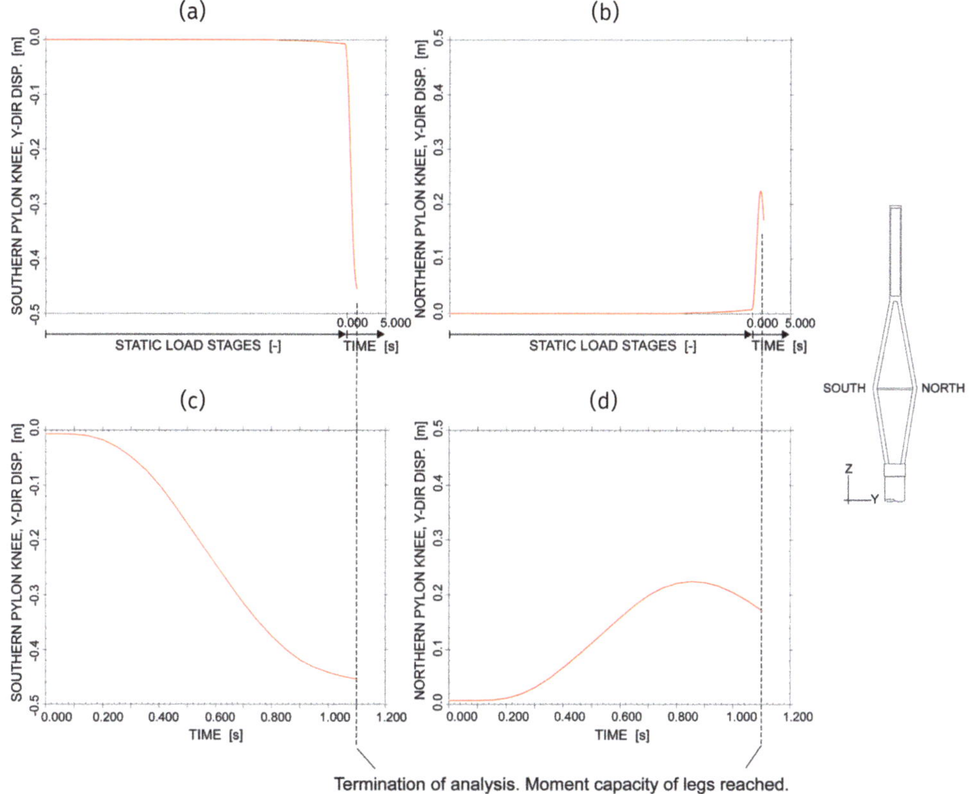

Termination of analysis. Moment capacity of legs reached.
Complete collapse of structure ensues

Fig. 9.14 Lateral displacement of pylon knees as function of time
a) Full displacement history of southern pylon knee
b) Full displacement history of northern pylon knee
c) Displacement history of southern pylon knee after rebar rupturing in diaphragm at reference time 0.000 s
d) Displacement history of northern pylon knee after rebar rupture in diaphragm at reference time 0.000 s

initiation of collapse are of limited magnitude compared to the displacements following the collapse. All of which occurs over a duration of ~1 s.

The complete collapse follows from very large, combined sectional forces at the base of the pylon and plastic hinge formation. At this stage, equilibrium cannot be sustained, which is identified by a dashed line in *Fig 9.14*.

While the knees move apart, the pylon head moves downwards. As the girder is vertically supported by the link slab and diaphragm, the stays closest to the pylon will decrease in tension. The two closest southern main- and side-span cables (T1BS, T2BS, T1AS, T2AS) decreases to zero stress, as can be seen directly in *Fig 9.15*.

Simultaneously with the downwards movement, a shift of the pylon head towards the main span occurs, evident in *Fig 9.17* and *9.18*.

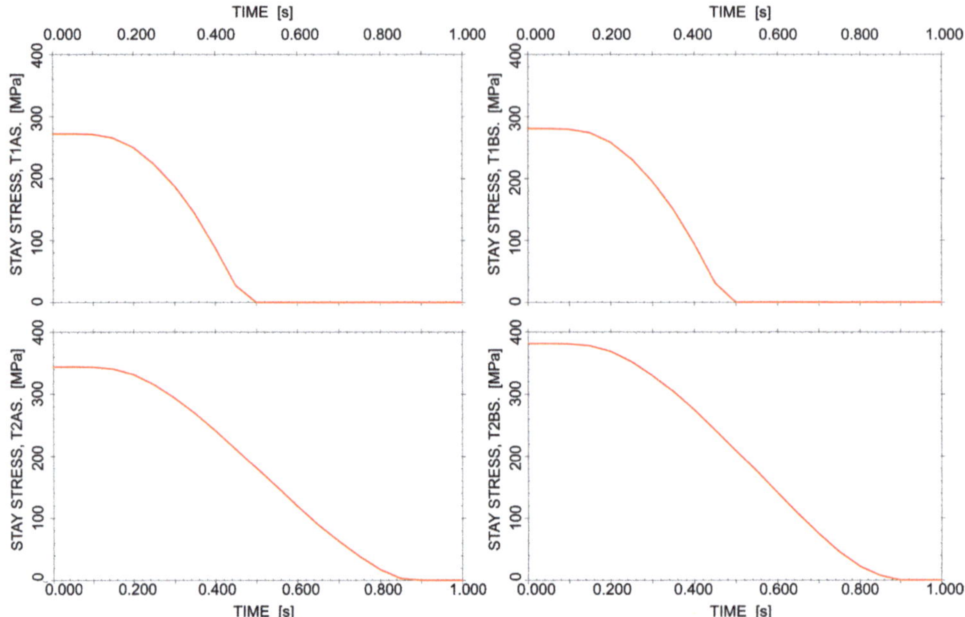

Fig. 9.15   Stay forces of south stays closest to main span as function of time after diaphragm rebar rupture

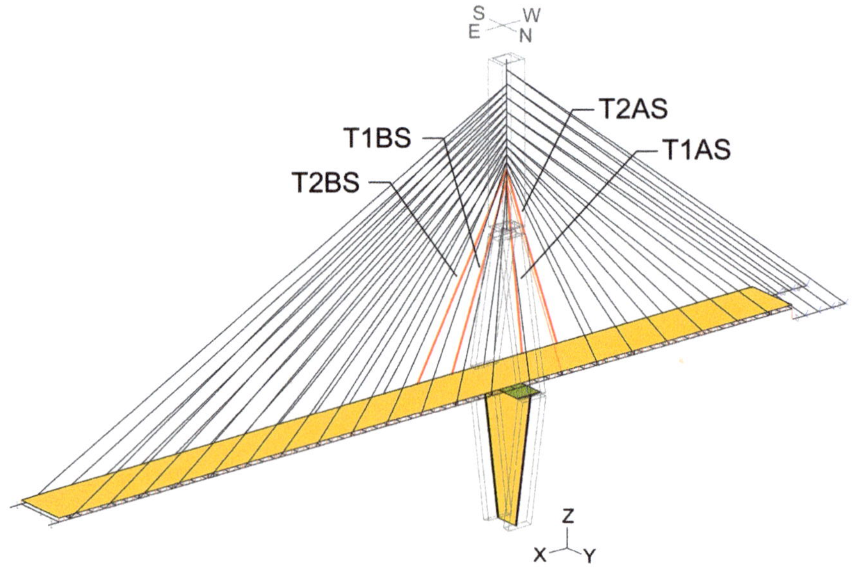

Fig. 9.16   Identification of stay cables with zero tensile stresses during collapse (see Fig 9.15)

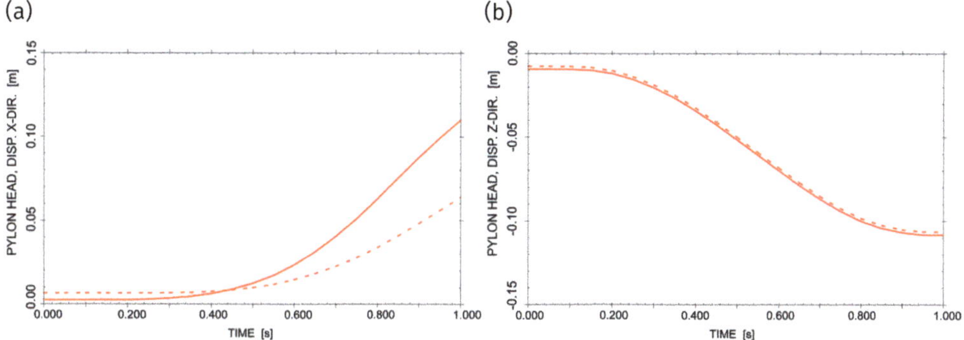

Fig. 9.17   Displacements of pylon head. Top of head (solid). Bottom of head (dashed)
            a) Displacement towards main span
            b) Downwards displacement

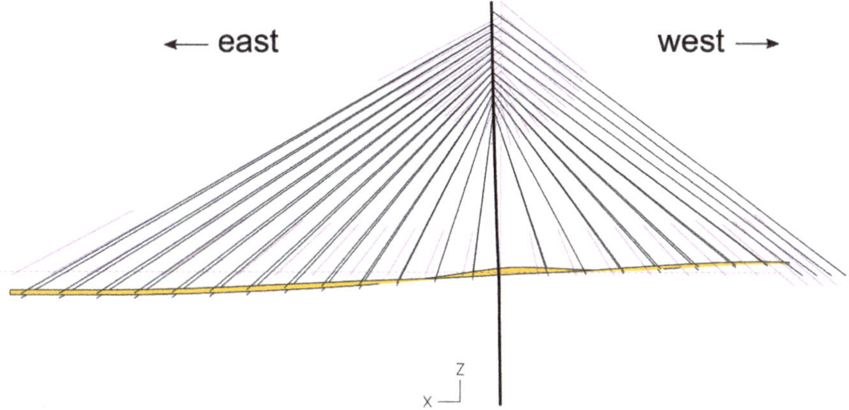

Fig. 9.18   Tilt of pylon top 1s into collapse shown with exaggerated displacements from
            FE-model

9.5.2 Comparison of Dynamic Failure Mechanism with Video

The team was provided with video footage of the collapse and 55 minutes leading up to the collapse [D10]. The video camera was situated next to a nearby road north of the bridge filming towards south. The video was filmed at ~15 frames per second with a resolution of 1280 × 720 pixels. The part of the view containing the bridge has a resolution of 285 × 285 pixels. The video was split into separate frames and then stabilized and corrected by image translation and rotation with the software Hugin to reduce wind-induced camera motions, which could otherwise be interpreted as incorrect movements of the bridge structure.

Fig 9.19 to *9.21* compares the observations from the video still frames with the FE-model results of the dynamic failure mechanism. Several similarities can be seen despite the low quality of the video.

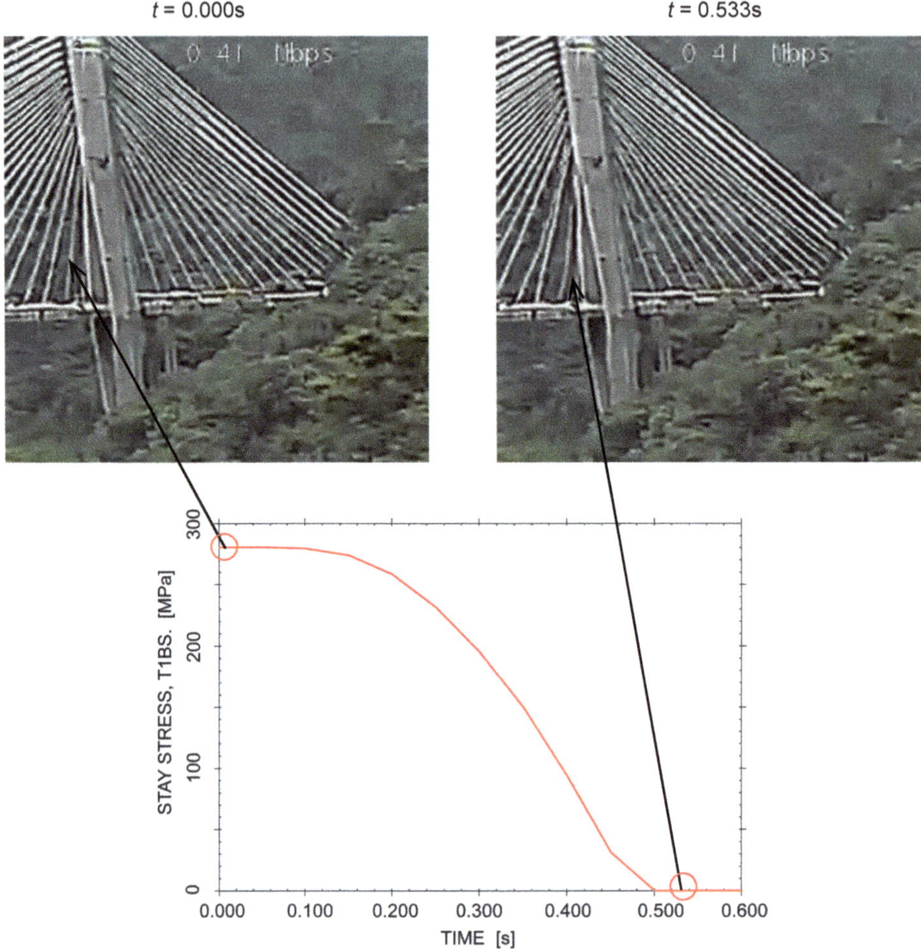

Fig. 9.19 Comparison of video and FE-analysis. Similar loss of tension in stay cables close to the pylon as observed in video are found from the FE-analysis

The collapse of Pylon B initiates on Jan 15, 2018, at 11:48:50 am. This is set as zero-time reference. The first visible sign of collapse is the slacking and loss of tension in the southern main-span stay cables closest to the pylon, clearly visible 0.5 s after the collapse has initiated. This is consistent with results from the FE-analysis, showing a loss of cable tension for southern stays closest to the pylon at 0.5 s after collapse initiates. At this stage, the south and north knees have moved outwards ~17 cm and ~11 cm respectively, which is unlikely to be visible from the low-quality video.

At ~1 s after initiation of collapse, a visible change of shading on the inside of the southern lower pylon leg is seen, indicating early separation between the pylon leg and diaphragm. This is consistent with the FE-analysis. Analysis results of nodal displacements show the southern knee moving faster and further away from the diaphragm than the northern knee. At ~1 s after the collapse, the southern knee has separated from the diaphragm by 45 cm.

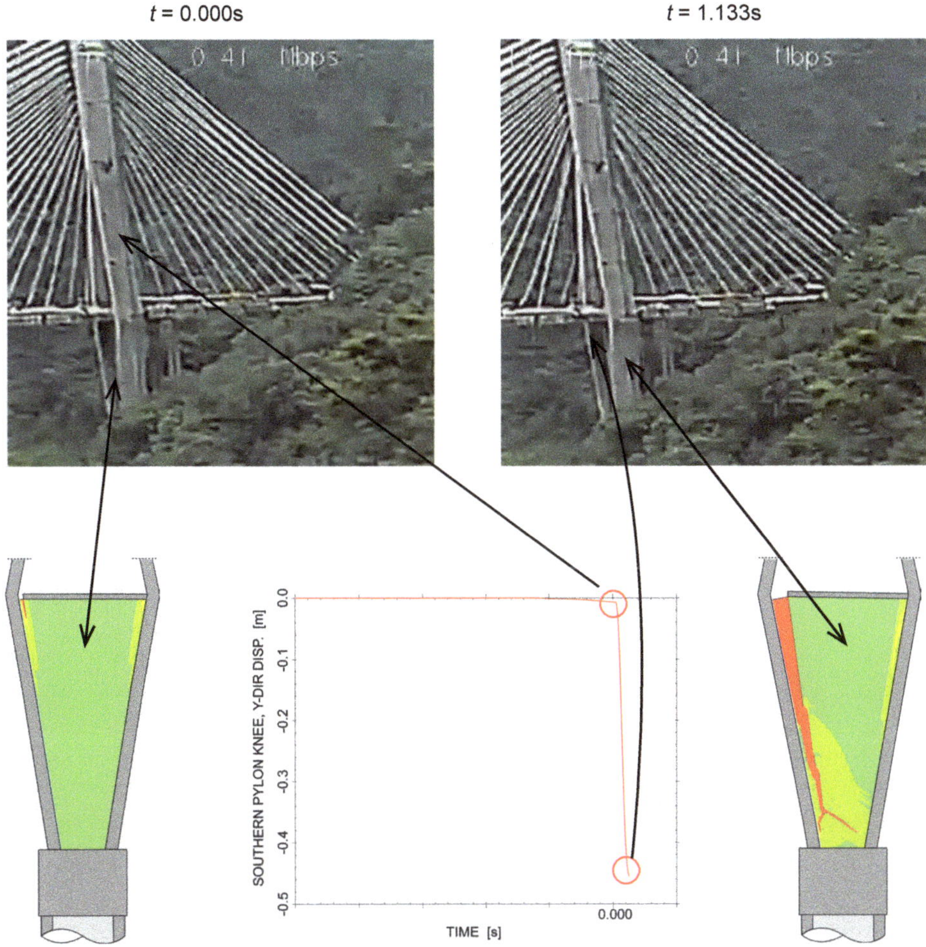

Fig. 9.20 Comparison of video and FE-analysis with respect to asymmetric separation of lower pylon legs and diaphragm. Same outwards movement of the southern knee is seen in both video and the numerical analysis

Results of diaphragm principal strains show values above the ultimate limit for the outermost diaphragm FE-elements, indicating an asymmetric separation allowing for no transfer of tensile stresses.

Following the separation of the lower pylon leg and diaphragm, the pylon head top displaces downwards and towards the main span. This is the result of the diamond shape expanding and the anchorage cables being fixed. Additionally, there is a downwards inclination of the main-span girder and slacking tension in side-span stay cables, as the side-span girder remains stationary.

The FE-analysis show results consistent with observations. The movement of the pylon head towards the main span is highlighted and compared with results from the FE-analysis.

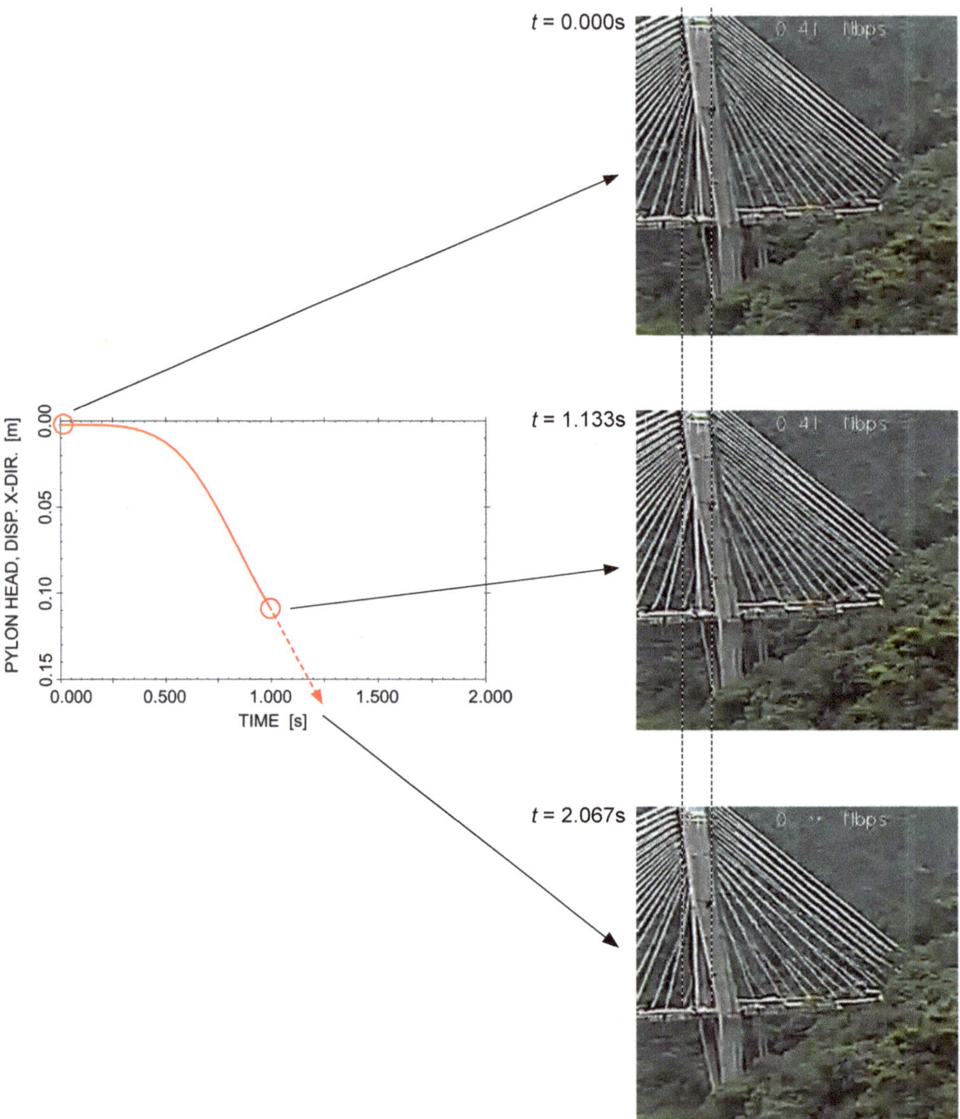

Fig. 9.21 Comparison of video and FE-analysis. The pylon head tilting towards the main span is seen from both the video and resulting from the FE-analysis

9.5.3 Global Reactions

The reactions immediately prior to collapse are summarized in *Table 9.6* for the support at the pylon base and for the abutment support, which can be identified in *Fig 9.2*. It is seen that a small lateral load of 3 kN is applied in the y-direction at the top of the pylon to reflect an asymmetric loading as described in Section 9.4.

Tab. 9.6   Summation of reactions at pylon base and abutment immediately prior to collapse

		Pylon base	Abutment
Force, X	[kN]	-392	392
Force, Y	[kN]	2.89	0.11
Force, Z	[kN]	69848	-3542
Moment, X	[kNm]	-26.8	0.00
Moment, Y	[kNm]	-13407	0.00
Moment, Z	[kNm]	0.50	0.00

9.5.4 Internal Forces, Pylon legs

The distribution of selected internal sectional forces in the pylon legs immediately prior to collapse is presented in *Fig 9.22*. The values of all sectional forces at top of pylon legs, knee level, and pylon base are presented in *Table 9.7*.

A slightly increasing axial force is found along the pylon legs due to the structural deadload of the legs themselves (see *Fig 9.22a*). The shear force in the upper pylon legs, found in *Fig 9.22b*, is introduced due to the structural deadload of the legs as well. However, it is seen that relatively large sectional shear forces are found in the lower pylon legs. This indicates that a significant part of the horizontal deviation forces at the pylon knees are absorbed by the diaphragm. This redistribution of horizontal forces into the diaphragm implies bending in the pylon legs (see *Fig 9.22c*).

Tab. 9.7   Sectional forces immediately prior to collapse at diamond top, knee level, middle of lower leg, and base of pylon.

		Diamond top		Knee level				Lower leg level		Pylon Base	
		South	North	South		North		South	North	South	North
				upper	lower	upper	lower				
Axial Force, R	[MN]	-16.87	-16.84	-24.86	-24.18	-24.83	-24.15	-26.09	-26.01	-31.20	-31.12
Shear Force, S	[MN]	-0.20	0.20	-0.20	-0.20	0.20	0.20	-0.20	0.20	-0.19	0.19
Shear Force, T	[MN]	0.41	0.41	-1.09	4.86	-1.09	4.85	0.07	-0.03	0.58	0.57
Rotational Moment, R	[MNm]	0.06	-0.06	0.06	-0.19	-0.06	0.19	-0.19	0.19	-0.16	0.16
Bending Moment, S	[MNm]	-0.46	-0.40	11.87	11.84	11.82	11.78	-7.34	-7.39	-0.90	-0.88
Bending Moment, T	[MNm]	6.18	-6.18	-0.74	-0.72	0.74	0.72	-2.35	2.35	-6.75	6.75

By evaluation of the results of the sectional forces in the pylon legs, presented in *Table 9.7*, it is evident that neither axial force, shear, nor flexural bending moments in the pylon legs were critical at the stage immediately prior to collapse. The axial force at the base of the pylon results in a low distributed axial stress due to the large cross-sectional area, see eq. (9.8).

$$\sigma_N = \frac{31.2MN}{A_c} = 3.3MPa \tag{9.8}$$

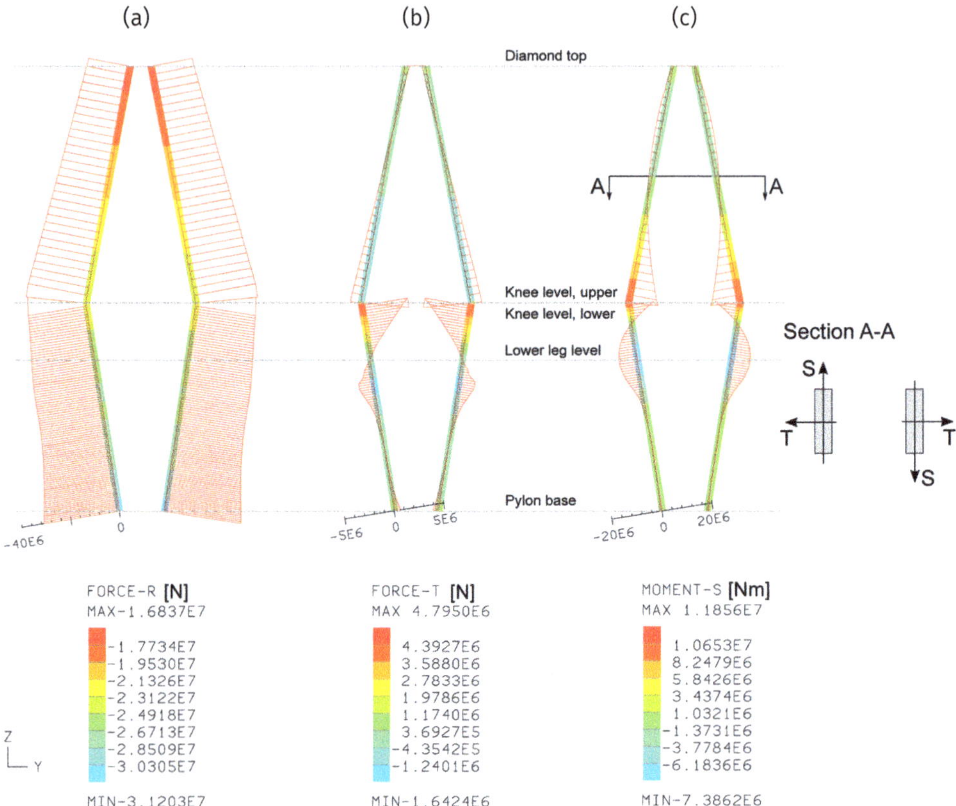

Fig. 9.22 Distribution of internal sectional forces in pylon legs immediately prior to collapse
a) Axial force
b) Shear force
c) Bending moment around weak axis

Axial force combined with two-directional bending is evaluated in the axial-force/bending interaction diagram of the pylon leg, presented in *Fig 9.23*. Here, all combinations of axial force and bending immediately prior to collapse from *Table 9.7* are shown. It can be identified that all combinations are allowed for even zero axial force, thus, the pylon legs were not initially the critical part of the collapsed structure.

The low utilization of the pylon legs immediately before collapse is verified directly in the FE-model which allows for direct evaluation of discrete rebar stresses. A 0.05‰ tensile strain and corresponding stress of 10 MPa is found in rebars of the upper southern pylon leg immediately below the knee.

The shear capacity of the pylon leg, ignoring contributions from concrete aggregate interlock and dowel action, was found to be 12 MN in Section 9.3.1. This corresponds to utilizations below 40 % and 70 % for the sectional forces prior to and after the diaphragm rebar rupturing respectively.

As stated in Section 9.5.1, the collapse is initiated by a sudden failure of the diaphragm, which implies a new equilibrium state of the structure. The abrupt change in internal

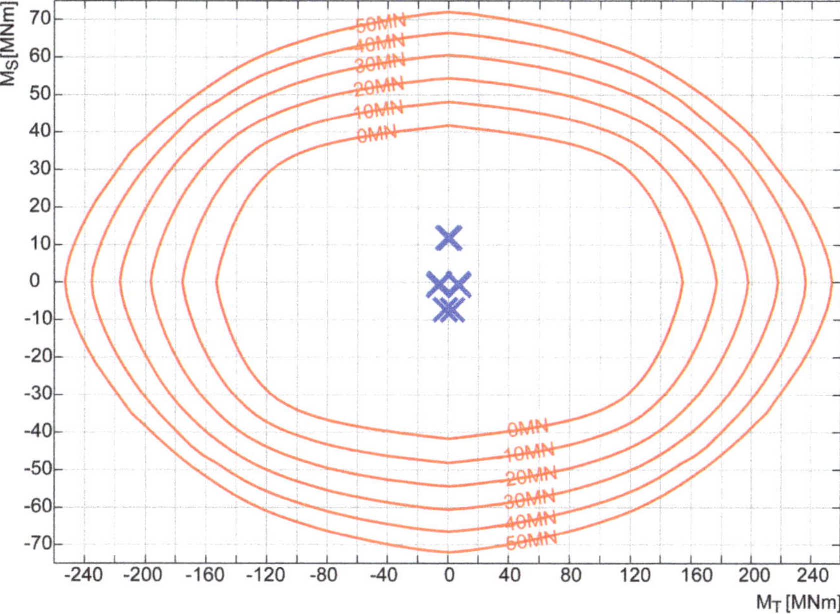

Fig. 9.23 Axial-force/bending interaction diagram of pylon leg together with combinations of sectional forces immediately prior to collapse

force path and equilibrium state results in large changes of pylon leg sectional forces compared to the stage immediately prior to collapse (see *Fig 9.24*). The sectional forces found from the dynamic analysis corresponding to 1 s after collapse initiation are found in *Table 9.8*.

As the southern leg separates from the diaphragm, twice as much axial force is carried by the northern leg. A slight increase in the total vertical reaction due to dynamic inertia forces is seen, as the combined reaction at the base is ~73.5 MN compared to the ~70 MN prior to collapse.

Plastic hinges have developed at both the south and north base with high moments around the pylon leg weak axis (M_s) of 59 MNm and 60 MNm respectively.

Tab. 9.8 Sectional forces after diaphragm rebar rupture at diamond top, knee level, and base of pylon

		Diamond top		Knee level				Pylon Base	
		South	North	South		North		South	North
				upper	lower	upper	lower		
Axial Force, R	[MN]	-14.73	-18.31	-23.35	-22.84	-26.80	-25.97	-29.18	-45.20
Shear Force, S	[MN]	-0.22	0.48	-0.22	-0.26	0.55	0.51	-0.31	0.12
Shear Force, T	[MN]	-1.73	-2.02	-2.67	2.81	-2.63	7.35	4.01	7.96
Rotational Moment, R	[MNm]	0.79	-0.39	0.71	-0.45	-0.18	0.98	-0.25	0.46
Bending Moment, S	[MNm]	-38.63	-42.03	49.27	49.01	51.86	51.81	-58.83	-59.85
Bending Moment, T	[MNm]	5.75	-5.85	-3.23	-3.25	12.99	12.78	-15.23	21.47

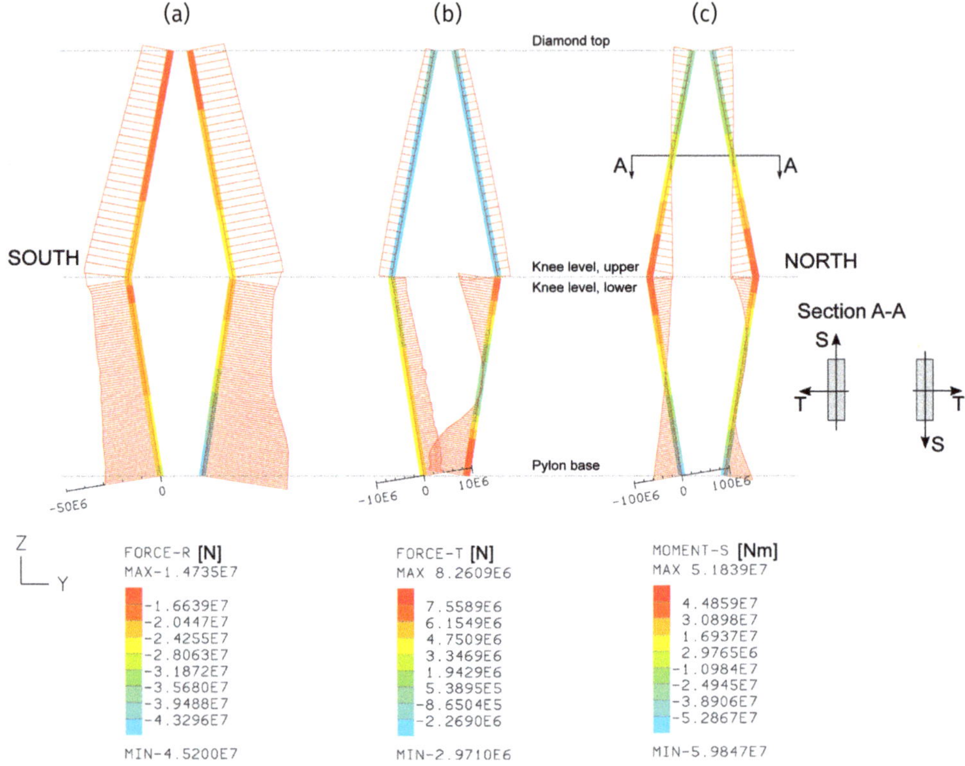

Fig. 9.24   Distribution of sectional forces 1s after diaphragm rebar rupture
a) Axial force
b) Shear force
c) Bending moment around weak axis

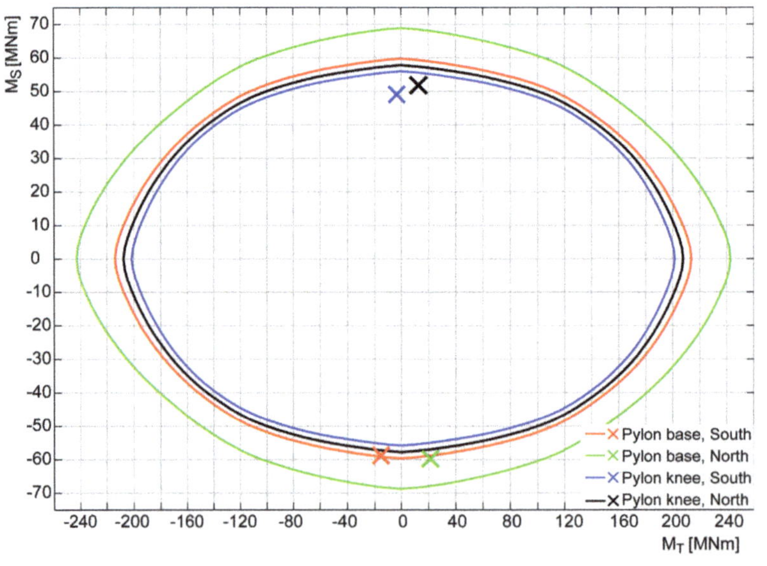

Fig. 9.25
Axial-force/bending interaction diagram of pylon leg together with combinations of sectional forces 1s after diaphragm rebar rupture

Combinations of the bending sectional forces at the base and knee level of the pylon together with the corresponding failure surface are seen in *Fig 9.25*. Especially the southern base sees utilization close to the capacity in combined axial force, shear and bending with $N = -29\,MN$, $V = 4\,MN$, $M_S = 59\,MNm$ and $MT = 15\,MNm$. And so, the dynamic analysis terminates, since no equilibrium can be found.

It should be noted that failure surfaces shown in *Fig 9.25* were calculated considering the actual material parameters found from material sample testing, including effects of yielding and strain hardening of reinforcement.

9.5.5 Stresses and Strains, Tendons and Link Slab

While the post-tensioned link slab is under compression in the transverse direction, following post-tensioning of the internal tendons, it absorbs a significant part of the horizontal deviation forces at the pylon knee, due to the high stiffness of this element. However, at loads approximately equal to the structural deadload of the pylon itself, compression in the link slab is lost as the knees move apart. When compression in the link slab is no longer sustained, the stiffness of this horizontal bracing element decreases substantially as it is then only governed by the relatively low stiffness of the 12 transverse tendons and no reinforcement connects the link slab with the pylon legs. A rough estimate of the relation between bracing stiffness of link slab with and without compression is given in eq. (9.9).

$$\frac{k_{comp.}}{k_{tendons}} = \frac{A_c E_c}{A_p E_p} = \frac{600mm \cdot 6000mm \cdot 30GPa}{12 \cdot 140mm^2 \cdot 190GPa} \approx 340 \tag{9.9}$$

Following this large decrease in stiffness of the bracing link slab, the primary restraint against the knees moving outwards is the presence of the diaphragm, accompanied by the flexural stiffness of the pylon legs themselves.

As the transverse tendons are introduced in the FE-model they exhibit a strain of 0.71 % and stress of 1362 MPa (19.4 t). At loads present immediately prior to collapse the strains have increased to a maximum of 0.81 % and corresponding stress of 1543 MPa (22 t) for the tendon closest to the main span. This is equivalent to 83 % of the guaranteed ultimate tensile strength, f_{GUTS}, which is high considering that loads are well below expected service loads. However, this indicates that the tendons were still in the elastic range immediately prior to the collapse.

As the pylon legs knees move apart to find a new equilibrium state after rupturing of the diaphragm rebars, the strain in the transverse tendons increases proportionally. Due to out-of-plane bending and torsional moments in the pylon legs, a slight difference is seen in the strain from the tendons closest to the main span and side span respectively. The transverse tendon closest to the main span exhibits ~5 % elongation at the time where the analysis terminates (see *Fig 9.26*). This falls between the 3.5 % guaranteed ultimate strain and the 6–7 % ultimate strain measured from test samples.

When evaluating these results, one should have in mind that the ultimate strain is often associated with a high standard deviation. Hence, even though the material tests

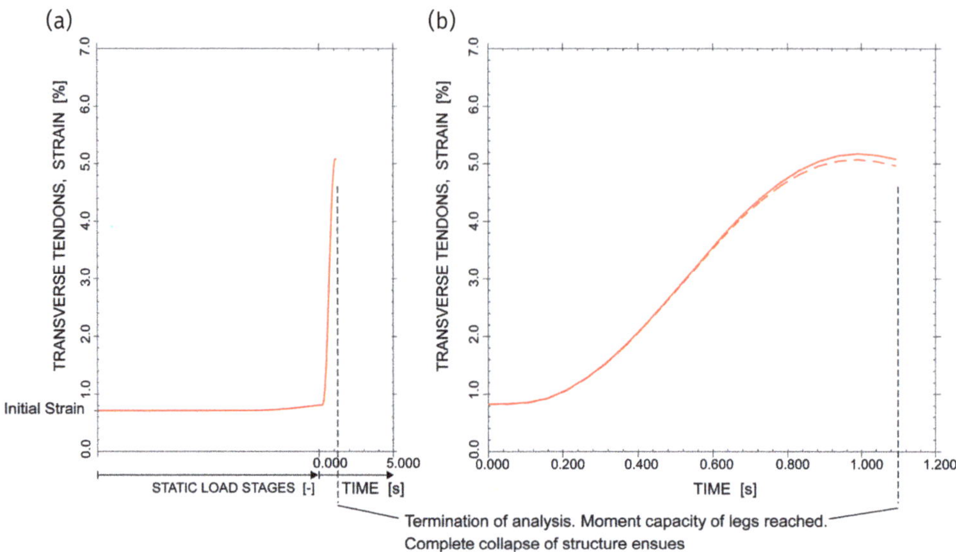

Fig. 9.26 Strain in transverse tendons as function of time. Tendon closest to main span (solid) and tendon closest to side span (dashed)
a) Full strain history of tendon
b) Strain history after diaphragm rebar rupture at reference time 0.000 s

indicate a slightly higher ultimate strain than the maximum found from the FE-model results, it is still likely that one or more of the tendons were not able to reach this level. If just one single strand of the 12 provided had a slightly lower ultimate strain, this would lead to the rupture of all 12 strands. Add to this that the material tests do not consider the provided grouting. Even though the grouting was very poor, it would probably have a slight restraining effect on the elongation of the strands, resulting in a lower average ultimate strain. Furthermore, the strands would have experienced a pinching effect where they exit the link slab as the pylon knee moves apart from diaphragm and link slab.

9.5.6 Stresses and Strains, Diaphragm

Redistribution of the horizontal deviation forces into the diaphragm implies that the upper part of this component acts as a reinforced concrete tension member. As seen in *Fig 9.27*, the zone of this upper part in tension extends concurrently with increasing load to achieve equilibrium.

As seen in *Fig 9.28*, the stiffness of the middle part of the upper diaphragm is significantly higher than the outermost part of the diaphragm, as the middle part is restrained by the link slab. This concentrates the strains in the upper diaphragm corners, amplified by the bottle-neck stresses due to the lack of re-bar connection between the link slab and pylon (see *Fig 9.29*).

The maximum allowable strain over the full width of the diaphragm is evaluated in Section 9.3.3 to be ~1‰ which corresponds to a total elongation of ~14 mm. This limit

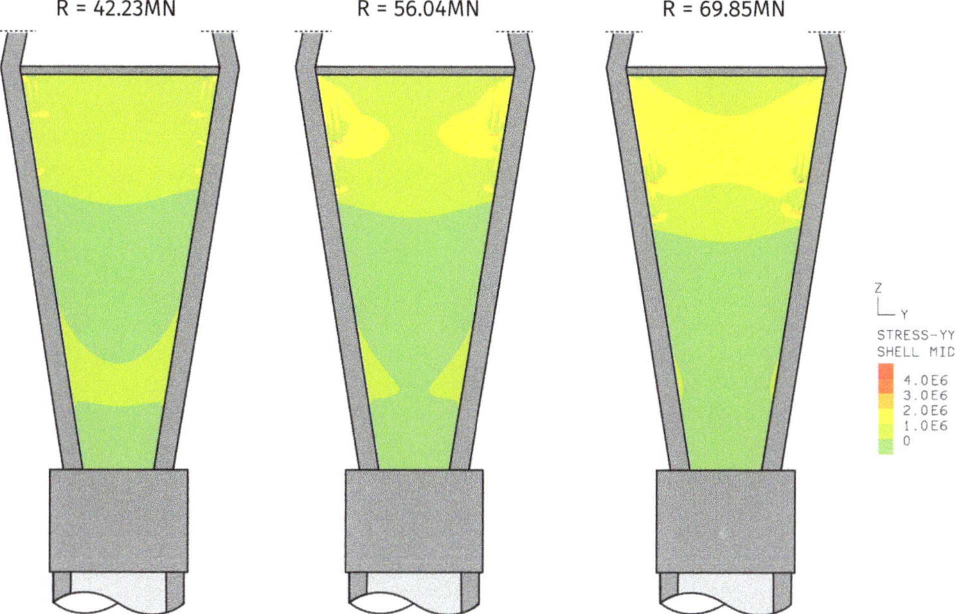

Fig. 9.27 Horizontal stresses in diaphragm, shown at 60 %, 80 %, and 100 % of the load present at the time of collapse

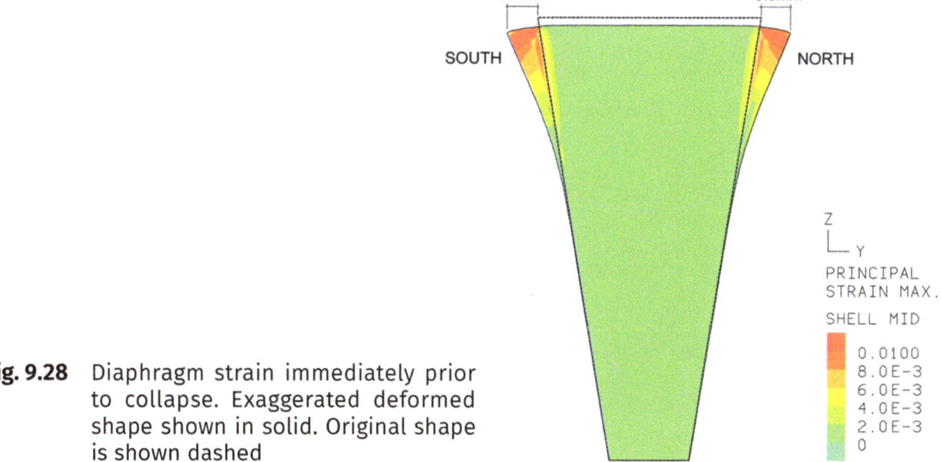

Fig. 9.28 Diaphragm strain immediately prior to collapse. Exaggerated deformed shape shown in solid. Original shape is shown dashed

is reached at ~101 % of the estimated load present at the time of collapse. This stage is presented in *Fig 9.28*, where an outward displacement in the y-direction of ~7.1 mm and ~6.9 mm are found at the southern and northern upper corners of the diaphragm. As already stated in Section 9.5.1, the strain limit is reached in the southern upper corner of the diaphragm, resulting in a sudden failure line along the southern pylon leg (see *Fig 9.30*). Beyond this point, the diaphragm does not contribute to equilibrating the horizontal deviation forces anymore.

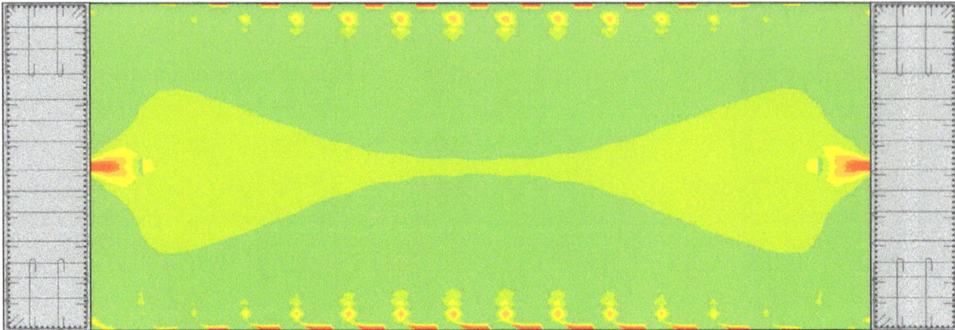

Fig. 9.29 Link slab stresses transverse to bridge axis. Concentrated "bottle-neck" stresses at the connection between the diaphragm and pylon. Tensile splitting stresses seen at the edge of the link slab

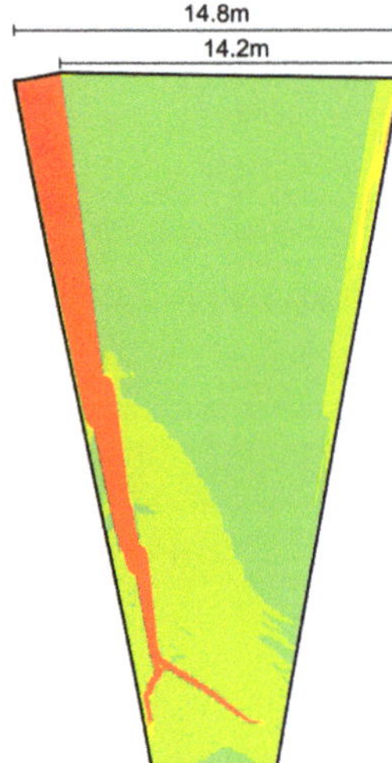

Fig. 9.30 Fully developed failure in diaphragm ~1s after exceedance of ultimate limit strain in upper southern corner

9.6 Summary of the Collapse

The dynamic failure mechanism found in FE-model results can be described sequentially as follows:

1. The construction process led to progressively increasing vertical loads in the pylon that translated into corresponding increasing lateral forces at the pylon knees. The lateral post-tensioning (tie capacity) with apparently low or no bond to the link slab was insufficient to counter the lateral (deviation) forces, and thus maintain compression in the unreinforced contact area of the pylon leg. This substantially reduced the stiffness of the structure and led to cracking and engagement of the uppermost connecting reinforcement at the interface between the pylon leg and diaphragm, as the lower legs need to deform to obtain a new equilibrium state.

2. The lateral connecting reinforcement between the pylon and diaphragm ruptured in the upper corner of the diaphragm, once the external load led to deformations that exceeded the local ultimate deformation capacity. The external load reached the level necessary for the collapse three days after the last major construction activity, i.e., the pouring of a concrete deck segment, and can be attributed to any number of normally inconsequential factors or their combination. These include, but are not limited to, creep, wind, temperature effects, and bearing movements.

3. After rupturing of the upper lateral reinforcement, the connection between the diaphragm and the lower pylon legs failed in a non-ductile asymmetric matter, so that the south pylon leg initially detached from the diaphragm in an "unzipping" of the lateral reinforcement through sequential rupturing of the whole diaphragm in less than one second (see *Fig 9.12*). Results of the FE-model show that the asymmetry of the collapse could have been readily caused by e.g., any small but normal geometric imperfections, like an acceptable deviation of the pylon from verticality, or a relatively small wind load.

4. The sudden, non-ductile failure of the diaphragm led to the dynamic amplification of the vertical forces acting on the pylon, which directly translated into an increase in the lateral forces at the knee. This led to increased lateral displacements of the pylon legs (see *Fig 9.14a* and *9.14b*.) with semi-plastic rotations at the knees and the base, whilst at the same time, the moment capacity requirements increased due to the large displacements, which increased the angular deviation of the upper and lower legs of the pylon and thus increased the lateral tie requirement. All of which resulted in an accelerated collapse.

5. While the knees moved apart, the pylon head displaced downwards and tilted towards the main span (see *Fig 9.1.7* and *9.18*). As the girder was vertically supported by the link slab and diaphragm, the stays closest to the pylon decreased in tension. The two closest southern main and side-span cables (T1A/BS, T2A/BS) lost tension (see *Fig 9.15* and *9.16*).

6. The pylon legs reached their moment capacity with large, combined sectional forces and plastic hinge formation at the base, as a result of the dynamic loads, and they were no longer able to support the resulting lateral displacements. At this stage, equilibrium could not be sustained (identified by the dashed line in *Fig 9.14*).

7. Further lateral displacements of the pylon legs eventually led to rupturing of the twelve steel strands that were placed laterally through the link slab at the location of the pylon knee.

8. Total collapse of the pylon and superstructure thus ensued.

Chapter 10

Pylon C - East

Apart from the depth of the foundation caissons, Pylon B and Pylon C were identical in their structural design and configuration. Furthermore, the superstructure supported by Pylon C was slightly behind Pylon B in construction, leading to an approximately 82 ton load deficit for Pylon C in relation to Pylon B (see Section 3.2.1). An inspection of standing Pylon C revealed the existence of two discrete vertical cracks in the diaphragm, starting from the uppermost corners, as can be seen in the photos in *Fig 10.1*. From this, it is seen that Pylon C reacted in the expected manner to the progressively increasing construction loads, i.e., through concentrated cracking of the diaphragm in the uppermost corners and, as a consequence, concentrated strains in the connecting reinforcement between the diaphragm and the pylon legs. This is described in this report as the first stage of collapse of Pylon B. Therefore, it is reasonable to conclude that Pylon C was on the verge of collapse before it was demolished. As its load deficit in relation to Pylon B was the main reason for it not collapsing before its demolition, any additional loading due to, e.g., working activities, vehicles, equipment, extreme temperatures, wind, or seismic actions, could have led to a collapse of the pylon, in a similarly catastrophic manner to what was observed on Pylon B and again without warning.

This does not imply that Pylon C could not have been hypothetically salvaged. Having identified the bridge deficiencies, appropriate strengthening and retrofit of the pylon, cables, girder, anchorages, and abutments could have yielded a serviceable part of a new bridge. What should be made clear though is that the risk to human life and the poten-

Fig. 10.1 Pylon C discrete vertical cracks in the uppermost corners of the diaphragm

tial cost implications related to a complicated and most likely risky repair and retrofit, would have likely rendered an attempt to salvage Pylon C prohibitive. It was due to these considerations that the expert team recommended that Pylon C with the associated superstructure be demolished using means that involved the lowest possible risk to human life. The expert team was informed that the demolition of Pylon C was carried out on July 11, 2018.

Chapter 11

Project Organization

The international panel of experts commissioned Leonhardt, Andrä und Partner, Germany, to undertake a review of the available documents, detailing the project organization behind the Chirajara Bridge. Key findings and observations are presented herein.

11.1 Reviewed Documents

Several documents have been reviewed, most of which were written in Spanish. However, the outcome of this review has been conducted in English, aiming to keep the meaning and philosophy of the original Spanish wording. Key findings and observations were based on the following supplied documents:

- Design and Build Contract (D&B) no. 123-OT-032-005 between CONINVIAL and GISAICO
- Specifications for Construction
- Regulations of Construction Works
- Control and Inspection Plan during Construction
- Phase II Design
- Minimum Personnel
- Quality Management System
- Plan for Technical Control
- Minutes of Follow-Up meetings during construction

11.2 Organization and Contractual Background

The Chirajara Bridge project was a part of a large national infrastructure plan, involving several parties. The parties involved, and their relationships, have been examined. In addition, an organizational chart summarizing these relationships is shown in *Fig 11.1*.

Agencia Nacional de la Infraestructura (ANI): Public Authority of the Colombian Government, Owner of the infrastructure. ANI subcontracted COVIANDES to operate several sectors, part of a large infrastructure plan, including a highway between Bogotá and Villavicencio.

COVIANDES was responsible for the Operation and Maintenance of the sector including the Chirajara Bridge project. COVIANDES subcontracted CONINVIAL to build and upgrade an existing segment of the highway between Bogotá and Villavicencio.

CONINVIAL was the Main Contractor, responsible for the construction works of the whole sector. CONINVIAL subcontracted other parties to build the key bridges within this sector. One of these bridges was the Chirajara Bridge, which collapsed in January 2018.

TRADECO was the initial contractor, contracted to design and build the Chirajara Bridge by CONINVIAL. The design work of TRADECO was referred to as the Phase II design work. The contract between CONINVIAL and TRADECO was terminated in 2016 before it could be finished.

GISAICO was appointed by CONINVIAL to take over the design and construction work, replacing TRADECO. Other parties were subcontracted by GISAICO to deal with some of the work related to the design and construction of the Chirajara Bridge. For instance, the authorized agent of VSL in Colombia was contracted for the supply of the stays. In the same manner, AREA INGENIEROS was hired for the Phase III design work.

AREA INGENIEROS was the Design Office, hired by GISAICO, to carry out the Phase III design works.

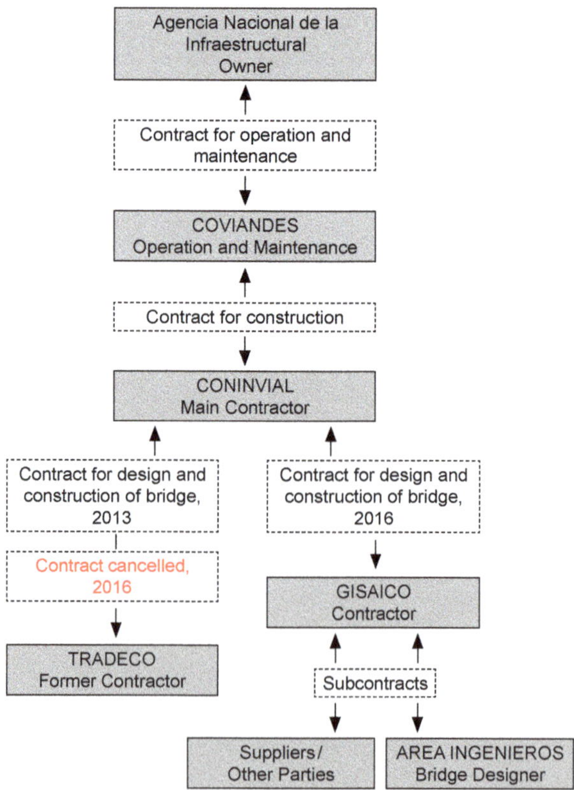

Fig. 11.1   Organization chart of the Chirajara Bridge project

A Design and Build Contract (D&B) was signed between CONINVIAL and GISAICO to clarify tasks, liability, etc. for the Phase III design work. The following key tasks and definitions of liability were included in this contract:

- GISAICO assumed all the risks associated with previous designs and executed works, in particular the foundation works of the bridge which were partly constructed by TRADECO and finished by CONINVIAL.
- Acceptance of the Phase II studies and design works of the Chirajara Bridge and application of these as a reference design.
- Execution of Phase III Studies and Design Works of the Chirajara Bridge according to the agreed Time Schedule (presented in Section 11.4).
- Procedures for changes in the Design
- Construction of associated Works within the agreed time schedules
- Nomination of key personnel for this project.
- Implementation of a Quality Management System with a Technical Control Plan.

11.3 Phase II Design

Phase II Design was the initial design documentation of former contractor TRADECO, provided to GISAICO by CONINVIAL, which included around 90 drawings of the Chirajara Bridge. In the D&B contract, it was stated that the Phase II design should be considered as a reference design for the Phase III design works, which were to be used for the construction of the bridge.

In Phase II, the bearing layout consisted of two anchor blocks, one located at each abutment. Four anchor cables on either side of the girder were anchored directly into the anchor blocks, featuring a fixed earth anchoring solution. The bridge girder was supported by the pylon and the anchor block at each abutment.

The pylon concept, defined in Phase II drawings, followed a classic diamond shape concept with two hollow sectioned pylon legs and a tie beam located at the knee level. The Phase II design specified substantial post-tensioning in the tie beam, linking the two legs. The post-tensioning cables consisted of 8 tendons comprising 19T15 strands each (see *Fig 11.2*). In addition, conventional tensile reinforcement was anchored into the pylon legs. The pylons were supported by mono caissons.

Stay cables were anchored in the pylon by concrete blisters at the inner faces of the pylon head. Post-tensioning bars were located at each face of the pylon to deal with the spreading forces introduced from the concentrated stay force in the pylon. Also, local reinforcement was specified in the blisters to cope with the bursting and spalling forces and to firmly tie the blister to the pylon wall.

Only a few drawings related to the girder were found in the Phase II design work. A girder construction comprising two longitudinal steel edge beams, connected by transverse cross beams including shear studs embedded in a concrete slab on top, was identified. The bottom flange of the crossbeam extended fully to the web of the longitudinal edge beams. The web of the longitudinal edge beams was designed with two longitudinal stiffeners.

(a)

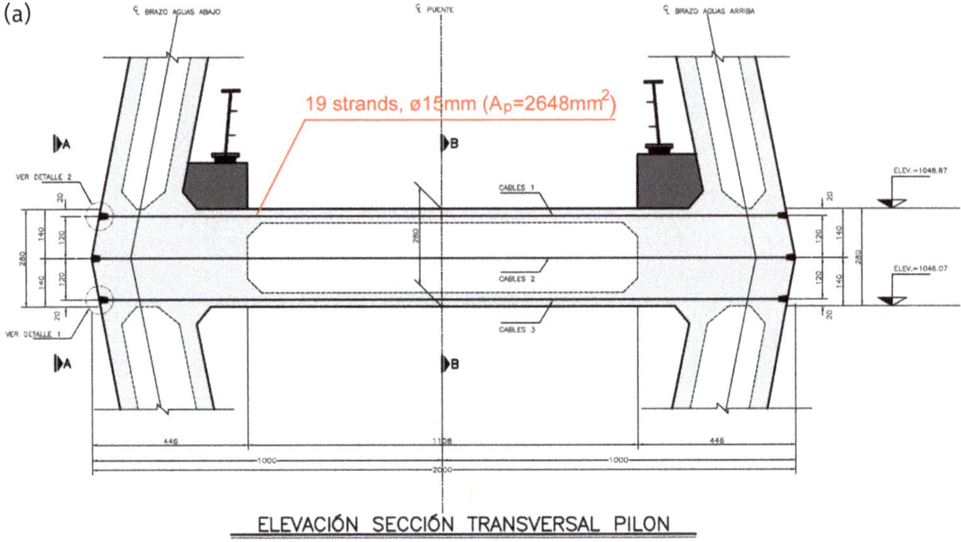

(b)

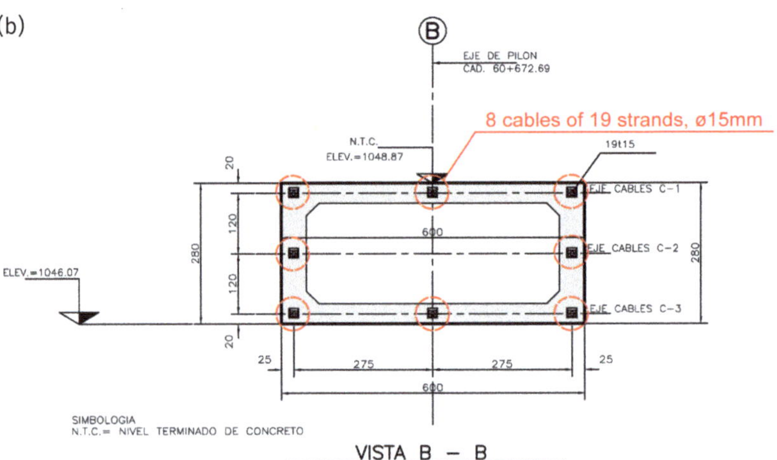

Fig. 11.2 Pylon details at knee level from Phase II design work
   a) Transverse section, knee level
   b) Longitudinal section, tie beam

Stay anchor forces are transmitted through a plate aligned with the web of the longitudinal girder, a so-called "web extension", a rather standard solution for steel composite stay-cable bridges. The location of the anchors seems to match the location of the crossbeam.

11.4 Phase III Time Schedule

The contract between CONINVIAL and GISAICO established a set schedule of submission deadlines for the Phase III design. The deadlines for submission of the design of specific structural components were:

- Caissons and abutments within 15 days following the signing of the contract.
- Pylon legs and diaphragm wall within 45 days following the signing of the contract.
- Pylons within 60 days following the signing of the contract.
- Homologated system of stay cables within 90 days following the signing of the contract

The complete Phase III design should be submitted from GISAICO to CONINVIAL within two months following the signing of the contract.

11.5 Quality Management System

The contract between CONINVIAL and GISAICO specified that GISAICO were responsible for implementing a Quality Management System in compliance with international standards, relating to the whole project. Internal and external audits were required by the contract.

Internally, GISAICO should exercise self-control and guarantee the quality of their own works and products through a Technical Control Plan and registers demonstrating this.

The Phase III design was carried out by the subcontracted design office AREA INGENIEROS, headed by engineer Héctor Urrego. The internal control of the design was carried out by the Director of Bridges following the internal procedure P-EO-01 of GISAICO. The details of procedure P-EO-01 were not found in the material provided.

11.6 Minutes of Follow-Up Meetings

Weekly meetings were held and attended by the different parties involved in the project, e.g., AREA INGENIEROS, GISAICO, CONINVIAL, SEC-VSL, and the external auditor. The minutes of these meetings reveal that the parties were aware of the changes that were implemented with the Phase III design and that deadlines were discussed in these meetings.

On several occasions, the meetings dealt with wind influence on the girder, especially how to keep the girder stable during construction.

The external auditor, represented by Germán García, prepared and kept a list of anomalies of the project. These included:

- Anomaly no. 082: Changes in the design have been carried out without the support of any technical justification.
- Anomalies no. 115 and 170: Incomplete information in the engineer design report. Lack of detailed analysis of the application of the loads, the structural analysis, and the evaluation of results.
- Anomaly no. 253: Grouting was not executed in the link slab of Pylon B immediately after stressing the tendons. The tendons were grouted several months after having been stressed. The time between post-tensioning and grouting of the ducts exceeded the limit specified in the INVIAS standard.

- Anomaly no. 347: Cracks were detected in the link slab of Pylon C
- Anomaly no. 348: The contractor planned on closing the main span of the bridge without the drawings and engineering design report being updated.

The auditor noted that GISAICO was not complying with the Technical Control Plan. The auditor repeatedly notified, with special emphasis, that GISAICO needed to document the changes and modifications of the design, along with submitting the technical reasons and calculations reports for approval by the client, before implementing any such changes.

11.7 Observations

Based on the material provided to the international panel of experts and the review of documentation conducted by Leonhardt, Ändra und Partner, the following observations have been made:

- It is not clear how the selection of the contractors TRADECO and GISAICO and the subcontracted design consultant AREA INGENERIOS was made. It has not been possible to find information elaborating on how the design consultant's qualifications and experience with major cable-stayed bridges were assured.
- By signing the Design and Build contract, GISAICO took all risks and assumed all liability for the past design and construction works of the bridge, carried out by the former contractor TRADECO. In a project of this size and complexity, it is very unusual to accept and assume all liability for the works and construction carried out by another party, even in the case where a thorough analysis and review have been conducted.
- Phase II design contained around 90 main drawings of the Chirajara Bridge, which covered in a good level of detail the abutments and the pylon, compared to the 57 main drawings that were provided as the Phase III design. Though incomplete, the Phase II design was detailed in the developed items and seemed to contain a higher level of quality and detailing for the pylon design compared to the Phase III design.
- The Phase III design did not contain any execution or erections drawings with associated method statements. Some drawings showed the stressing load per tendon and the number of stay cable strands. Nevertheless, the construction sequence was not identified nor described.
- Changes from the Phase II reference design to the Phase III design introduced significant changes to the structural concept relating to bearings, stay anchorages, and pylon design. Specifically, the pylon design was changed from a typical diamond shape with a tie beam between the pylon leg knees to a design featuring a link slab at the knees and a diaphragm wall between the lower pylon legs.
- The amount of post-tensioning and tie capacity at the pylon leg knees was reduced by a factor of ~13 from the Phase II to Phase III design.
- The Time Schedule for submission of the Phase III design seems unrealistically tight for preparing a complete and adequate design with the required level of detail

for a structure of this complexity unless significant parts of the design had already been prepared before the start of the Time Schedule.

- As the contract explicitly mentions the diaphragm wall, which was not included in the Phase II design, it is possible that some parts of the Phase III design had already been established by GISAICO before signing the Design & Build contract with CONINVIAL. Yet, minutes from the follow-up meeting revealed that all parties seemed under significant pressure to comply with the submission schedule and set deadlines.
- Unrealistic time schedules may have driven the designer and contractor to skip elementary but necessary steps in the design and build process, with the consequence of proposing and executing a structure with a faulty structural concept.
- As required by the Design & Build contract, GISAICO implemented a Quality Management System and appointed internal auditors and supervisors. From the information provided, it was evident that quality control was executed for the materials used in the construction works.
- No records of internal checking of the Phase III design or any of the design changes during construction were provided, even though it was formally defined by the Quality Management System.
- The external auditor emphasized during the follow-up meetings that changes in the design were carried out without the support of any technical justification. Also, it was stated that both the design drawings and the engineering design report lacked information and needed to be updated, even as the contractor was approaching completion of the main span of the bridge.
- Several changes to the Phase III design were not properly documented or implemented in design drawings but were only communicated through the WhatsApp communication platform.
- No information was found clarifying whether the bridge should be considered a key risk and of an importance for which an external and independent Category III checking by own methods of analysis and review of design would ordinarily be required.

Chapter 12

Recommendations

12.1 General

The causes for the total collapse of the Chirajara Bridge, Pylon B, and its associated superstructure under construction have been extensively covered in the preceding sections of this report.

The review of the design used as the basis for the construction, the post-collapse observations on site, and subsequent re-analysis of the bridge structure as designed, as well as a Category III independent check have revealed that the design was inadequate for its purpose. For details refer to the *Executive Summary* and the other sections of the report.

Although the immediate cause for the collapse was identified to be insufficient strength of the horizontal tie between the "knees" of the pylon, the project was also found deficient in many other structural and organizational respects. It has been determined that the collapse of the bridge would have been always imminent after erection of the pylons and application of loads during construction or in service.

After this catastrophic event, it is important to try to understand how and why this catastrophe happened. How could such a fatal design error be committed, and why would it pass unnoticed in subsequent phases of the work? As it is most often the case, based on experience from previous investigations of for example crashes of transport vehicles and airplanes, it is rarely one single cause that leads to catastrophic failures, but rather a succession of events which individually could often be dealt with, or prevented from having any consequences, but collectively accumulate beyond ability to handle and eventually cause the inevitable crash.

For the current investigation it is, therefore, necessary to review and analyze the process for the construction of the bridge facility from the initial contracting of the concession by the Owner to the Operator and via the Construction contract to the Main Contractor to the Design and Build Contract for the Bridges, including subcontracts to the designer and suppliers. These comprise key elements of the overall description of the project intentions, including required criteria of safety, as well as operation and maintenance requirements throughout the life of the facility. The procurement of construction firms, procurement of design firms, handing over to the Operator with maintenance instructions and manuals, etc. as well as setting up a suitable operation, inspection and maintenance program for the bridge are key components to outline. Such a review has been performed. The key findings and observations from this review are presented in Section 11.

The entire chain of events including the important interface and communication of data and project requirements between the many parties involved needs analysis to identify the weak links and subsequently revise or establish new robust practices with adequate redundancy which will reduce risks of similar nature in the future and assure realization with a high degree of safety, also against errors and omissions in the sequence of events themselves.

The international industry involved in major infrastructure works has developed over many years to become rather robust and safe through free and open dissemination and exchange of knowledge and experience via appropriate international associations like Owner's fora, IABSE, FIDIC, NSPE, and many more.

Best practices have been developed over time and recommended practices and design codes have been documented by various international organizations and institutions; in particular the International Federation of Consulting Engineers, FIDIC.

The originally typically used procurement process in the Anglo-Saxon tradition was that the Owner or Operator would employ an independent Consulting Engineer, experienced in the type of construction envisaged. The Consulting Engineer would prepare designs in several consecutive stages from initial conceptual designs over preliminary design for discussion and decisions by the Owner and further to a fully detailed design, with detailed technical specifications and contractual conditions to be used for competitive tendering and engagement of a qualified contractor who would build the works. During the construction, the Consultant would supervise construction on behalf of the Owner, to ensure that the design intent is achieved and that the construction satisfies all specifications and fulfills the contractual conditions, as well as approve design changes as necessary. At the completion of construction, the Consultant would prepare As-built drawings as well as maintenance manuals and instructions for Operation and Maintenance during the lifetime of the structure. This process assures that the structure is properly designed, and construction supervised so as to comply with design intentions.

In the southern European tradition, the Owner or Operator would via qualified own staff go through the phases described including preparation of an outline design as a basis for a call for competitive design-build tenders for the Owner's design, including an invitation to submit alternative designs fulfilling the requirements and at the same time optimized for the experience and equipment of the individual contractors or consortia.

In recent decades, the two different methods have converged, and it has become an often-used practice to use a process of procurement that combines the best practices of the two alternative methods of procurement. The Owner/Operator employs a consulting Engineer to develop several conceptual designs satisfying the Owner's requirements for comparison of options, prices, aesthetics, time for construction including risk of delays as well as operation and maintenance matters. This process serves primarily to prepare the Owner/Operator for proper identification and definition of his requirements, and normally ends with his selection of a so-called "Illustrative Design" or "Employer's Design" to assist in focusing on the communication of the requirements to other parties.

In combination with detailed performance criteria, specifications, and conditions, the Illustrative Design forms the basis for the call for competitive design and construction bids from prequalified Contractors who would be required to associate with competent design consultants. The bids must satisfy and comply with the criteria expressed by the Specifications and the Illustrative Design as well as the contractual conditions.

The evaluation of the bids will be performed by the Owner and his Consultant after which a Design and Construction contract will be closed by the parties.

The advantage of this procedure is that the full responsibility for the works will be unequivocally placed with the Contractor and his design consultant. And the Owner/Operator would benefit from the latest technological development of the individual competing contractors in combination with designs optimized for these methods.

The method of procurement is fully described and documented in the FIDIC EPC/TURNKEY Contract [14] as a recommended practice, accepted by all parties in the industry.

12.2 Specific Recommendations

The primary cause of the collapse has been determined with great certainty together with several critical deficiencies, which did not directly contribute to the collapse but were related to the overall bridge design. The event would have been much less likely to occur if a reasonable Standard of Care had been exercised in the process of the realization of this bridge, as described in the preceding section.

Based on observations made throughout this report, particularly in Sections 2, 7, 9, and 11, it is recommended that:

- internationally published documents, recommendations, and standards, depicting procurement practices for the owners of infrastructure projects, are reviewed. An example is the FIDIC standards, which have been developed by experienced authors based on extensive experiences. These documents are widely accepted and used as good practice in the industry.
- a quality management system like the ISO 9000 system [15, 16] is implemented and followed. This will ensure that all participants in the projects satisfy a certain well-defined level of standards. The complete chain of involved parties needs to comply with such requirements from Owner/Operator of the facility to designer, contractor, subcontractors, vendors, and material and equipment suppliers, checkers, supervision teams, operation and maintenance teams, monitoring systems, etc.
- a realistic time schedule is outlined, accounting for unexpected delays and events. Do not accept contracts, which imply unrealistic submission times for the design.
- robust communication between all related parties is ensured. Do not rely on, for this purpose, inappropriate communication systems like WhatsApp, etc.
- a system of independent checking of any design and construction is implemented to maximize assurance that errors and omissions will be caught by third parties. This includes the requirement of an independent Category III design check for any structure of importance.

- an appropriate degree of structural robustness, redundancy, and ductility is ensured. The benefit of this is that if, against odds, the structure approaches collapse,
  this will be preceded by warning in form of e.g., excessive deformations which
  ensures ample warning and opportunity to instigate provisions that attenuates the
  consequences or even prevents collapse from taking place.

The strictest and systematic standards cannot alleviate problems if the personnel using
them are not qualified for their function. Quality is generated by qualified personnel.
Therefore, it is strongly recommended to instigate a personnel development and ed-
ucation program, for all parties in the process so they can live up to their individual
responsibilities for implementing new infrastructure projects of importance like complex
bridge structures with sufficient experience and background.

Chapter 13

Lessons Learned

Laurent Rus Jenni, Founder and CEO at Singular Structures Engineering

13.1 Preamble

Whenever a structural collapse occurs, engineers should take a moment to reflect on working methodologies from governance to operation and maintenance stage corresponding to the life cycle of the structure, including planning, design, commissioning, and construction stages. Moreover, if the information is shared by the involved parties with maximum transparency and collected within the shortest time possible, it is the whole of the structural engineering practice that benefits from a continuous learning process, contributing to the "state of the art" of civil engineering.

This report on the Chirajara bridge collapse has clearly shown several aspects of our bridge engineering practice that tend to be overlooked, either by pressure of time or budget, both aspects that jeopardize the safety of civil engineering projects entrusted to engineers by society.

This report highlights in great depth a) how the bridge design was flawed on several key structural elements, b) the lack of standard use of technical documents and procedures, and c) a non-robust contractual organization, as three basic elements which have led to the failure of the structure.

During the investigation process, the triggering effect of the Chirajara Bridge collapse was identified to have occurred during the construction stage where construction loads already exceeded the design load at the pylon knee with the corresponding detailing between tower leg, diaphragm, and link, which has low ductility, leading to a sudden collapse of the structure.

While the triggering effect and first location of structural failure have been described, the report also shows multiple potential locations for partial or total collapse and how these design flaws were neglected by engineers over the different stages of the project.

The availability and the openness during the interviews of the main contractor, design engineer, site engineer, and cable supplier and their acceptance to be interviewed by the independent investigation team, which was composed of structural engineering experts lead by Professor Christos T. Georgakis, have to be stressed. This openness should not be taken for granted but should be appreciated for the benefit of a constantly evolving practice of bridge design and construction.

Some of the lessons learned that can be extracted from this investigation of a singular structure are shown in, but not limited to, the following summary, which is structured corresponding to the life-cycle of a bridge. Note that the provided lists here-after are non-exhaustive given the wide range of project set-ups and local practices. However, these lists specify key technical documents and procedures that are based on good practice in the industry.

13.2 Governance Stage

One of the key elements of the governance stage that is highlighted in the Chirajara Bridge investigation report is the strategy of the contractual delivery and of the selection process of the contractor along with the subcontracted design consultant.

Guidance material at the governance Stage includes:

- **International Legal Standard Documentation FIDIC** (« *Fédération Internationale des Ingénieurs-Conseils* »)

  Over the years, FIDIC has constantly improved their documentation by adjusting, modifying, and adding new forms of contracts, including a default hierarchy of the different forms with a clear definition of key roles and responsibilities. The most significant FIDIC contracts for internationally built singular structures are:

  ○ FIDIC Engineering Procurement, and Construction (EPC)/Turnkey Contract (Silver Book)
  ○ FIDIC Design-Build and Operate (DBO) Contract (Gold Book)

- **International Quality Management System (QMS)** to comply with ISO Standards, specifically, and validated through good industry practice:

  ○ ISO 9001 Quality management systems (requirements),
  ○ ISO 19011 Guidelines for auditing management systems, and
  ○ ISO 9004 Managing for the sustained success of an organization – quality management approach.

- **Standard Technical Documentation**

  While recognizing that every single structure and project set-up are different from one another, there are still some good industry practice standard details that are the result of years of expertise, both in the research and analysis fields and in construction, which provide robustness to the structural solution proposed. Standard technical documentation covers norms, codes, manuals, guidelines, manufacturers' and suppliers' recommendations. The latter would have a wide range of sources, so that careful evaluation is necessary. Nevertheless, for singular structures such as a cable-stayed bridge, a recollection of technical documentation for previously constructed bridges is required:

  ○ Conventional structural detailing at cable deck and tower head connection;
  ○ Conventional longitudinal shear transfer detailing for deck composite section behavior;

- ○ Stay cable, bearings, and movement joints according to technical specifications;
- ○ Construction sequence strategy (stage analysis) with minimum and maximum target deflections at control points;

- **Financial Budget Allocation Strategy**

For public infrastructures, where the taxpayers' money is used, funds should be wisely invested (secured throughout by a government or state budget and an appropriate procurement process) for an efficient, resilient and robust infrastructure. Financial scenarios shall be developed and evaluated based on current and future market conditions. Combined mechanisms of loans, bonds, public and/or private finance are to be considered in combination with fund release periods (milestones payments). Milestone payments constitute an incentive to design-build (DB) contractors, when properly and clearly defined, by reducing the contractors' financial risks.

- **International Procurement Methods**

A number of procurement processes can be applied, from EPC to DBO processes (including DBOMaintain, and DBOFinance), the latter being the most commonly used procurement process for singular structures such as cable stayed bridges. A traditional DBO process would include a Request for Qualification (RFQ) to arrive at a shortlist of joint ventures (JV) composed of contractors (Main and Subcontractors), design consultant (Main and Support Designers) and operator (Operator), This shortlist of joint ventures would be invited to the next phase of the procurement process, the Request for Proposal (RFP) stage. During the RFP stage, the different JV teams develop the proposed bridge scheme design to a higher level of detail. The selection of the awarded JV DBO team is based on the assessment of several parameters, including, e.g., special integration, aesthetics, construction method, life cycle cost, construction time, environmental impact, use of local resources, or innovation. These criteria shall be clearly stated in the RFP design basis.

In parallel, an independent design consultancy firm, Owner's Engineer (OE) must assist the Bridge Owner (Client/Authority) with the technical validation of the design and construction strategy developed by the different JVs, both during the competition and DB stages.

- **Preliminary Risk Assessment and Corresponding Mitigation Measures**

This process shall be revisited and updated at the key milestones of the design life of the project, which is often avoided or misunderstood by most of the "hands-on" engineers. This reluctancy could result from the complexity of the whole process, i.e., a) identifying all potential risk scenarios, b) structuring those into a meaningful approach with clear identification of linked or related risks, c) clearly appointing a responsible person to the mitigation of each risk, and finally d) assessing the associated cost, either in time, material or resources, which is a highly subjective task.

While it is legally possible for a Main Contractor to subcontract most of the works, especially fora singular structure, to a subcontractor, and still maintain the contractual responsibility of the execution in relation to the Bridge Owner, this scenario should have generated some serious doubts on the Main Contractor's capability of successfully achieving the mandate. This became evident when the first contractor (TRADECO) was declared bankrupt at the end of 2016 with all the caissons (A to D) and anchor-block A completed (anchor-block D was partially completed at that time).

Indeed, the infrastructure delivery process should take into account local engineering and construction practices along with international good practices for a successful project delivery by using an appropriate framework of contract mechanisms, project time frames, and corresponding budget and project risk sharing among the parties involved.

It is important to select the optimum mechanism that would provide a transparent, robust and competitive approach for the ultimate goal of delivering a sustainable and resilient infrastructure for society.

13.3 Planning Stage

Guidance material at the planning stage includes:

– **Business Case**

A business case provides justification for undertaking a project, with an estimated program and cost. It evaluates the benefit of several alternatives, defines the risks, and is typically developed using the following key concepts:

  - Strategic context: The compelling case for infrastructure development;
  - Technical alternatives: Scenarios of infrastructure technical solutions, both land use and structural typologies;
  - Economic analysis: Return on investment based on investment options;
  - Financial approach: Derived from a state or regional sourcing strategy and/ or design/construction procurement process strategy for a given time frame;
  - Management approach: e.g., project stages, procurement process, roles and responsibilities (governance structure), life cycle choice, concession periods, or payment mechanisms;

– **Technical Preliminary Studies**

The preliminary technical evaluation of complex interventions in the public domain is essential but is frequently neglected as a result of a perceived urgent need, either from society or politicians. Preliminary studies, often termed feasibility studies, are a key step in the planning stage to validate some of the preliminary assumptions for a more robust, resilient design and construction process, thereby reducing costs and minimizing potential unknowns that would harm the implementation of the project. Such studies shall a) confirm assumptions, both technical and financial, b) confirm potential procurement processes, c) identify new potential risks and assign responsibilities, and d) provide recommendations for new studies.

A list of technical preliminary studies would be:

- o Topographical and bathymetry survey;
- o Geotechnical and hydrogeological site investigation and interpretative report;
- o Hydrological Studies - regional and river flow hydrology;
- o Natural site hazard studies;
- o Road and bridge general arrangements;
- o Traffic demand;
- o Special technical specification requirements;
- o Preliminary environmental impact assessment;

13.4 Design Stage

The design stage corresponds to the process stage of detailing to the construction level of the originally developed scheme design. This process goes through intermediate stages, starting from concept design to confirm the overall geometry of the structure (for each key structural element), to further implement the design criteria and design requirements for the construction, and also to define any additional studies and laboratory testing requirements (wind tunnel / seismic table / hydraulic testing, including structural detail testing/mock-ups for strength, ductility, fatigue, durability, constructability validation) in order to achieve a successful detailed design stage.

The design stage is often broken down into key structural components (foundations, tower/piers, abutments, deck, cable stay, pavement, etc.) and into several milestones (30 %, 60 %, 90 %, and 100 % design milestones – interim and final submissions) so that the ongoing design process can be reviewed by OE and Independent Engineers (IE), adjusted by the designer (as needed), and finally approved by the IE.

A design of 100 % completion (detailed design) should have a sufficient level of detail for the contractor to build the structure and should include information and guidance for the assumed construction process.

DB contracts are considered "fast track" contracts as the design and construction works overlap, as detail documentations for key structural components are approved. Guidance material at design stage (interim and final submissions) includes:

- Design brief and set of calculations;
- Relevant technical notes, studies and reports;
- Technical specifications;
- Special provisions;
- Drawings, including construction sequence;
- Bill of quantities and cost estimate
- Health and safety plan;
- Operation and maintenance plan, and corresponding;
- Review by an Independent Engineer – Category III Check;

The IE is typically contracted by the Bridge Owner (Client), in order to maintain independence in the integrity, stability and durability assessment of the structure and its compliance to project specific performance requirements, and therefore shall not be part of the awarded JV team, or any of its subsidiary firms.

The Category III check requires an independent assessment of the structure from design load definition, design load combinations, design calculations, numerical analysis, and review of the proposed construction sequence, among other checks.

The so-called *"first principal"* approach is a very important and effective tool, which consists in the breakdown of a complex problem into simpler basic elements to get an understanding of the key/primary structural behavior. The understanding of the structure can then be developed from this point in a stage-by-stage reasoning process. The basic element used in this report is the strut and tie model that breaks up all the internal forces (bending and torsional moments, shear and axial forces) into two types of forces, i.e., tension and compression forces, resulting in a truss model structure, which requires every truss node to be in equilibrium. The resulting truss forces can then be compared to the structural capacity of each section.

Chapter 2 "Assessment of Bridge Design" demonstrates that the essence of the first principal approach on the design of several key structural components was not met (e.g., transition beam, anchor beam, tower lower leg-diaphragm, tower head, cable to deck connection) and therefore the appropriate degree of robustness, redundancy and ductility were not provided.

In addition, it is worth highlighting the necessary consistency across all technical documentation that allows for cross-checking assumptions and detailing across several responsible entities, and the coherence in the provided design and construction documentation (refer to chapter 2.8 "Selected Drawings Inconsistency"). Furthermore, concordance between the developed numerical analysis model and the resultant structural detailing shown on the drawings is required. In other words, there is an engineering judgement in the definition of the connection degrees of freedom in the numerical model that needs to be consistent with the structural detailing (flow of forces and section equilibrium) shown on detail drawings for construction or assembly.

13.5 Construction Stage

It is good practice for the contractor to develop:

- **Erection Procedure (EP)**

  The EP includes erection sequence drawings, an assessment of erection stresses, any temporary supports, as well as details on camber/stay forces for geometry control (deformation and rotation) at each erection stage;

- **Construction Monitoring Plan (CMP)**

  The CMP includes the definitions of the necessary site equipment to monitor, e.g., vibrations, deflections, rotations, temperatures, or wind speed (location and frequency of measurements) with the corresponding target values and response threshold values. Should a threshold value be exceeded, the CMP shall include a communication procedure;

- **Method Statements (MS)**

  MS include equipment and material to be used, safety and environmental control measures, and a contingency plan in the event of any accident or damage. MS documents also include QA / QC references as hold points, witness points, testing and inspection milestones.

- **Inspection and Test Plans (ITP)**

  ITP include tests and inspections to be carried out for each constructed element, highlighting, e.g., the type of testing and the standard procedure to be followed, the number of sample tests, location of testing, and record format.

The report highlights the lack of design change mechanisms (change procedure), or at least, the inappropriate fulfilment of such procedure, had it existed. While changes in the design and during construction are perfectly understandable and certainly are part of the risk of fast-track projects, it is also true that a proper mechanism for a change procedure has to be established, both for changes issued from the authority (Bridge Owner / Client) and the Contractor. These may be significantly different procedures but they ultimately have the objective to document a possible change of a) cost (e.g., direct and indirect cost, impact of life cycle cost, proposed payment schedule), b) time schedule (e.g., duration, target dates, licensing, permits), and c) detail description and documentation of the change at design and construction stage.

The report disclosed several versions of cable stay forces received by the cable supplier, ultimately leading to a girder geometrical position higher than expected. Cable structures are flexible structures and highly dependent on the construction and load sequence to achieve their final geometrical position and thus loading. Any change in the cable force form the originally assumed condition requires a re-assessment, both structurally and geometrically, which was not performed (refer to chapter 11.7 "Observations"). In addition, it is a good practice that at the completion of each of the key structural works, a construction report including the following information be submitted to the bridge owner for record and future use, should it not already be stated in the construction contractual arrangement:

- Construction monitoring plan;
- Construction method statements as well as inspection and test plans;
- Records of the laboratory and field tests;
- Records of any remedy actions undertaken;
- Summary of the issues encountered during the construction of the structural element;
- Records of the temporary works left on-site;
- Record of compliance with structural design drawings and specifications documents;
- Records of non-compliance and corresponding review and approval processes;
- As-built drawings;

An interesting aspect of the communication process between Designer and Contractor, which is indicated in the report and highly relevant nowadays, is the use of WhatsApp (or similar communication platforms) for quick messaging exchange (voice and written message) without all necessary validation across disciplines and peer engineers and also without the required level of record for instructions or design changes. Needless to say, that the use of such a communication platform is not banned from our day-to-day professional work. However, even if it is very useful, engineers have to be aware that this tool does not relieve from using the proper communication channels, procedure, and forms as part of the quality management system.

13.6 Operation and Maintenance

This section has been developed for consistency and completeness of this chapter, organized following the different stages of the life cycle of a structure.

The term Operation and Maintenance (O&M) shall be understood in its wider sense in the following context as it also includes responsibilities such as Inspection (I) and Repair (R), which could be handled by a joint venture of its own.

Guidance material at the Operation and Maintenance (O&M) stage includes:

- Design brief (by Designer);
- Project structural (and non-structural) element breakdown (by Designer);
- Long term planning activities and cost planning (by designer and O&M organization);
- O&M drawings (O&M organization);
- O&M monitoring plan (O&M organization);
- Blank report template (by Contractor);
- Inspection report template (by Contractor);
- Deformation report template (by Contractor);
- O&M manual (O&M organization, and validated by the bridge owner), i.e., health and safety plan, environmental management, quality assurance and quality control, inspection and maintenance plan (IMP - O&M organization), and inspection procedure (IP - O&M organization)

From the above suggested list, it is important to note the hand-over procedure from the Contractor to the O&M organization, whereby the Contractor develops a blank report, an inspection report and a deformation report to set the base line for future O&M works to prevent potential claims:

- To record any previous inspection works performed, if applicable, and its correlation to the as-built drawings;
- To gather instrumentation, measurement procedures and records, should this be available;
- To capture the specific deformation profile at several key points in order to produce a stage documentation corresponding to the recorded position at hand-over time in order to determine any long-term deformation;

13.7 Closing Remarks

After the dust has settled following the collapse of a structure, there is always the feeling that this failure could have been avoided. While mistakes can occur, these shall be prevented by control mechanisms to reduce any risk to the bare minimum at the time of construction and continuously over the life cycle of the structure. The underlying condition of the collapse of the Chirajara Bridge has been the lack of exercising the duty of care as required in the design and construction of an unconventional cable stay bridge, as well as a lack of quality assurance and quality control over the different stages of the project.

This report highlights the importance of, among other aspects:

- Professional skills (specifically educated and trained professionals);
- Access to technical information;
- Access to standard contracts;
- Access to standard procedures for quality assurance and quality control (document and communication management system);
- Change Procedure;
- Communication Procedure.

Following the collapse of a structure, the first steps after the failure are essential and privileged for data and evidence gathering. The first visit to the site of the collapsed Chirajara bridge from the team of experts was on March 20th, 2018, barely two months after the bridge collapse. While this may seem a long time since the collapse, this period includes the contractual arrangements, expert team mobilization, pre-assessment of readily available information, travel arrangements, and setting a site visit agenda, all of which was necessary to prepare for the in-situ investigation. In this regard and in view of a continuous and improved learning process, a task group could be dedicated to developing a process and/or a team to support bridge owners in such unexpected and catastrophic events.

An important issue immediately after such a failure is the question of responsibilities. Who is responsible for

a) the rescue of individuals;
b) public safety, including the structural stability of the wreckage, and;
c) preserving the site and defining and developing records for the collection of evidence?

The preservation of the site and an early record of the perishable evidences (evidences that could deteriorate by environmental exposure) becomes crucial. This includes the documentation of the final position of the collapsed structural elements which would allow validating the collapse mechanism of the edifice (refer to chapter 4.2 "Orientation and Placement of Debris").

Finally, some additional questions to reflect on when a structural collapse occurs: has the project organization influenced the ultimate result? Has the project procurement process influenced the ultimate result? What appears to be the major causes of structural collapse in terms of number of collapses? In the last 20 years, have there been any changes in codes, standards or practices following a structural collapse?

References

[1] Fernández Ruiz, M. Muttoni, A., Gambarova, P. G., *Analytical Modeling of the Pre- and Postyield Behavior of Bond in Reinforced Concrete.* Journal of Structural Engineering, Vol. 133, No. 10, 2007, pp. 1364–1372. DOI: https://doi.org/10.1061/(ASCE)0733-9445(2007)133:10(1364).

[2] Beeby, A. W. *Ductility in reinforced concrete: why is it needed and how is it achieved?* The Structural Engineer, Vol. 75, No. 18, 1997, pp. 311–318.

[3] Marti, P., Alvarez, M., Kaufmann, W., Sigrist, V. *Tension chord model for structural concrete.* Structural Engineering International, Vol. 8, No. 4, 1998, pp. 287–298. DOI: https://doi.org/10.2749/101686698780488875

[4] Ožbolt, J., Bruckner, M. *Minimum reinforcement required for RC beams.* European Structural Integrity Society, Vol 24, 1999, pp. 181–201. DOI: https://doi.org/10.1016/S1566-1369(99)80069-7

[5] Seguirant, J. S., Brice, R., Khaleghi, B. *Making sense of minimum flexural reinforcement requirements for reinforced concrete members.* PCI Journal, Vol. 55, No. 3, 2010, pp. 64–85. DOI: https://doi.org/10.15554/pcij.06012010.64.85

[6] Ghosh, S. K. *Minimum reinforcement requirements to prevent abrupt flexural failure of prestressed concrete immediately following cracking.* ACI Structural Journal, Vol. 84, No. 1, 1987, pp. 40–43. DOI: https://doi.org/10.14359/2778

[7] Goto, Y. *Cracks Formed in Concrete Around Deformed Tension Bars.* ACI Journal Proceedings, Vol. 68, No. 4, 1971, pp. 244–251. DOI: https://doi.org/10.14359/11325

[8] European Committee for Standardization (*CEN*). *EN-1992-1-1:2014, Eurocode 2: Design of concrete structures*, 2004.

[9] European Committee for Standardization (*CEN*). *EN-1993-1-5:2006, Eurocode 3: Design of steel structures – Plated structural elements*, 2006.

[10] AASHTO, *LRFD Bridge Design Specifications*, 7. ed, 2014

[11] AASHTO, *ASTM A615 -15a – Standard Specification for Deformed and Plain Carbon-Steel Bars for Concrete Reinforcement*, 2015.

[12] Niels J. Gimsing, Christos T. Georgakis. *Cable Supported Bridges: Concept and Design (Third Edition).* John Wiley & Sons, Ltd, 2012. DOI: https://doi.org/10.1002/9781119978237

[13] The International Federation for Structural Concrete (*fib*). *Model Code 2010*, 2012

[14] International Federation of Consulting Engineers (*FIDIC*). *Conditions of Contract for EPC Turnkey Projects (2017 Silver Book)*, 2. ed, 2017

[15] International Organization for Standardization (*ISO*). *ISO 9001:2015, Quality management systems – Requirements*, 5. ed, 2015

[16] International Organization for Standardization (*ISO*). *ISO 9000:2015, Quality management systems – Fundamentals and vocabulary*, 4. ed, 2015.

Notation

Throughout the report, all unspecified units are defined according to the SI system.

ε	=	strain
ε_c	=	strain at maximum concrete compressive stress
ε_{ct}	=	maximum concrete tensile strain
ε_{cu}	=	ultimate concrete compressive strain
ε_s	=	steel strain
ε_y	=	yield strain
ε_u	=	ultimate strain
ε_{bu}	=	bond ultimate strain
ε_{su}	=	ultimate steel strain
$\varepsilon_{RC,u,sing.}$	=	ultimate strain of reinforced concrete member with single crack formation
$\varepsilon_{RC,u,dist.}$	=	ultimate strain of reinforced concrete member with distributed crack formation
f_c	=	concrete cylindrical compressive strength
f_{cu}	=	concrete stress at maximum compressive strain
f_{ct}	=	concrete tensile strength
f_y	=	steel yield stress
f_{su}	–	ultimate steel stress
$f_{y,eff}$	=	yield stress, effective cross section
$f_{u,eff}$	=	ultimate stress, effective cross section
f_{GUTS}	=	guaranteed ultimate tensile stress
σ	=	stress
σ_s	=	steel stress
σ_c	=	concrete stress
σ_N	=	axial stress
τ	=	bond stress
$\tau_{b,max}$	=	maximum bond stress

δ	=	relative bar-concrete slip
A_2^*	=	torsional flutter derivative
A_s	=	reinforcement area
A_c	=	concrete area
A_p	=	strand area
$\varnothing_s$	=	reinforcement bar diameter
F	=	force
R	=	support reaction
N	=	axial force
V	=	shear force
V_s	=	shear capacity
M	=	moment
L	=	length
E	=	modulus of elasticity
E_s	=	modulus of elasticity, Steel reinforcement
E_p	=	modulus of elasticity, steel strands
E_0	=	initial modulus of elasticity, concrete
E_h	=	hardening modulus, Steel reinforcement
ν	=	Poisson's ratio
γ	=	specific weight
λ	=	local punching coefficient
θ	=	concrete shear crack inclination
s	=	spacing
s_{rm}	=	mean crack distance
l_p	=	plasticized length of bonded bar
k	=	spring stiffness
$\rho_{s,l}$	=	longitudinal reinforcement ratio
$\rho_{s,H}$	=	horizontal reinforcement ratio
$\rho_{s,V}$	=	vertical reinforcement ratio

The Investigators

Christos T. Georgakis
Aarhus, Aug 31, 2018

Yozo Fujino
Yokohama, Aug 31, 2018

Siegfried Hopf
Stuttgart, Aug 31, 2018

Klaus H. Ostenfeld
Copenhagen, Aug 31, 2018

S. Eilif Svensson
Copenhagen, Aug 31, 2018

www.ingramcontent.com/pod-product-compliance
Lightning Source LLC
La Vergne TN
LVHW071527180726
843512LV00014B/1188